RIVERSIDE REGIONAL
LIBRARY, BOX 389
JACKSON, MO 63755

The Proper Care of
REPTILES

With time, patience, and attention, you can induce your reptiles to breed in captivity and produce many young, like this attractive baby Black Racer, *Coluber constrictor constrictor*. Photo by John Dommers.

JOHN COBORN
TW-115

Facing page: The Reeve's Turtle, *Chinemys reevesi*, is an Asian species that has only recently enjoyed commercial popularity. Photo by J. Visser.

© 1993 by T.F.H. Publication, Inc.

Distributed in the UNITED STATES to the Pet Trade by T.F.H. Publications, Inc., One T.F.H. Plaza, Neptune City, NJ 07753; distributed in the UNITED STATES to the Bookstore and Library Trade by National Book Network, Inc. 4720 Boston Way, Lanham MD 20706; in CANADA to the Pet Trade by H & L Pet Supplies Inc., 27 Kingston Crescent, Kitchener, Ontario N2B 2T6; Rolf C. Hagen Ltd., 3225 Sartelon Street, Montreal 382 Quebec; in CANADA to the Book Trade by Macmillan of Canada (A Division of Canada Publishing Corporation), 164 Commander Boulevard, Agincourt, Ontario M1S 3C7; in ENGLAND by T.F.H. Publications, PO Box 15, Waterlooville PO7 6BQ; in AUSTRALIA AND THE SOUTH PACIFIC by T.F.H. (Australia), Pty. Ltd., Box 149, Brookvale 2100 N.S.W., Australia; in NEW ZEALAND by Brooklands Aquarium Ltd. 5 McGiven Drive, New Plymouth, RD1 New Zealand; in the PHILIPPINES by Bio-Research, 5 Lippay Street, San Lorenzo Village, Makati, Rizal; in SOUTH AFRICA by Multipet Pty. Ltd., P.O. Box 35347, Northway, 4065, South Africa. Published by T.F.H. Publications, Inc. Manufactured in the United States of America by T.F.H. Publications, Inc.

The Proper Care of REPTILES

JOHN COBORN

Another popular pet species is this Common Tegu, *Tupinambis teguixin*. Photo by Isabelle Francais.

CONTENTS

Section One—Care and Husbandry

Introduction 7
Facts About Reptiles 15
Classification 32
Housing 40
General Management 67
Health and Hygiene 90
Captive Breeding 103

Section Two—A Selection of Species

Order Chelonia 123
Order Crocodilia 140
Order Rhynchocephalia 142
Order Squamata 144

Index 252

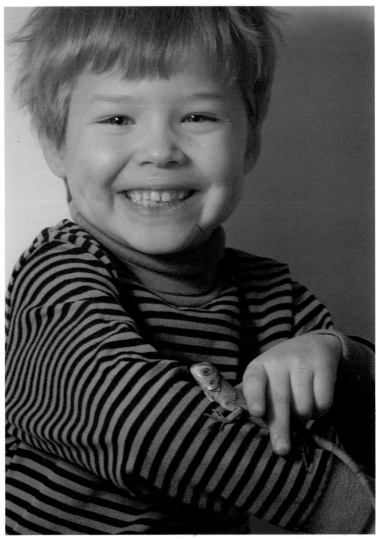

People of all ages can appreciate the appeal of reptiles. Photo by Isabelle Francais.

SECTION ONE—CARE AND HUSBANDRY
Introduction

"Reptile" is the familiar collective name given to a member of the class Reptilia, which includes the turtles, tortoises and terrapins; the crocodiles and alligators; the tuatara; the lizards; the amphisbaenians; and the snakes. The class Reptilia is a medium-sized group of vertebrates (with about 7,000 described species) that is often included together with the class Amphibia (frogs and toads, salamanders and newts) in the study of herpetology. Herpetology (derived from the Greek *herpeton*, a creeping thing) is a branch of zoology that deals with the evolution, classification, biology, and distribution of two vertebrate classes, the Amphibia and the Reptilia.

Reptiles bring out a whole range of reactions in people. As they are a relatively little known and poorly understood group of animals, the majority either profess to "hating" them or, at best, to regarding them as just another kind of "nuisance" that the world would be much better off without. How often do we hear the question "What good are snakes?" or "What good do alligators do?" A person with any degree of intelligence would, of course, not need to ask any of these questions, but would already have answered them. A good response to such a question, however, could be another question: "What use are humans?". And if we take a look around the world as it is today, that question is very easy to answer.

One of the most pertinent statements regarding man's attitudes toward reptiles (and amphibians) was made at the beginning of the present century when the herpetologist Hans Gadow wrote in the preface to his *Amphibia and Reptiles* (1901): "One reason for the fact that this branch of Natural History (herpetology), is not very popular, is a prejudice against creatures which are clammy and cold to the touch, and some of them may be poisonous. People who delight in keeping Newts or Frogs, Tortoises or Snakes, are, as a rule, considered eccentric. But in reality these cold-blooded creatures are of fascinating interest provided they are studied properly."

In more recent times herpetologists, both professional and amateur, are found in relatively greater numbers and are, indeed, on the increase. Improvement in communications through the media has undoubtedly played a major part in this phenomenon. Today, amateur herpetologists are more readily accepted and are considered to be almost as "normal" as water skiers, big-game hunters, or stamp collectors!

There are a great many herpetologists who have a special love for certain groups of reptiles, and there are those who delight in keeping turtles, lizards, or snakes. Others have a more general interest, perhaps studying reptiles in the wild, photographing them, or keeping examples of all groups for a time in order to observe their habits more

INTRODUCTION 9

Displayed here is a very attractive tank that can be used for almost any small reptile. Photo by Susan C. Miller and Hugh Miller.

INTRODUCTION

A garter snake (*Thamnophis* sp.) is one of the many excellent starter pets. Photo by John Dommers.

closely. An expanding popular interest in wildlife, probably impassioned by the excellent, ever-improving nature films to be seen on TV, has led to an increasing awareness of the need to conserve all species and not just the "cuddly" ones. It is perhaps ironic to say that some of the more vulnerable reptile species may only be saved from extinction by breeding them in a protected or captive environment.

Considerable improvements in our knowledge of the captive husbandry and breeding of the reptiles have occurred in the last two or three decades. Although some species have always been reasonably easy to keep alive, successful breeding was a different matter. Other species have formerly been classed as "difficult" or "impossible" to keep. It was the general realization that most reptiles are relatively "non-adaptive" to

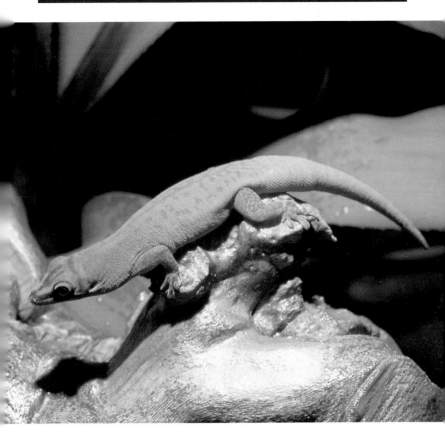

This is the attractive but seldom seen day gecko, *Phelusuma astriata*. Photo by K. T. Nemuras.

environmental changes that vastly improved our chances of not only keeping them alive and healthy but also propagating them. When a reptile is removed from the

INTRODUCTION

wild it must be kept in an environment very similar to that from which it originated, and this must include the appropriate seasonal climatic changes. It is therefore very important to have a detailed knowledge of the natural habitat of a species and its reproductive habits before any attempt is made to keep it in the terrarium.

Those who keep reptiles

A python of this size would probably not be happy in a small cage like the one shown. Photo by Susan C. Miller and Hugh Miller.

INTRODUCTION

in the home terrarium have an extremely fascinating hobby that has several advantages over some of the more conventional mammal or bird pets. Relatively little space is required for a terrarium that can be esthetically pleasing, attractive, and an instant point of conversation whenever visitors arrive. Terraria can be set up complete with plants and animals, thus bringing a little nature into the lives of even those living in city high-rise apartments. Given a few simple requirements, reptiles are clean, odor-free, quiet, and non-demanding. Once the basics have been set up, maintenance and cleaning chores are minimal, and you have every opportunity to do your own herpetological research.

In the following text I have compiled information primarily for the the prospective home terrarium keeper, but I have also included material that will be of interest to all people concerned with reptiles both in the wild and in captivity. The early chapters deal with general information and care, while later chapters briefly describe a selection of species from the reptilian orders. This section is not intended to be a taxonomic reference or even a field guide to help enable the reader to identify species—that is outside the scope of this book. Information on positive identification should be gleaned from a good field guide to the appropriate area (when available) and by reference to the scientific literature and major herpetological journals in college libraries. The present text is intended purely to demonstrate the infinite variety of shapes, patterns, colors, and habits that occurs in the reptile world, and to present some guidelines on the care of some individuals or groups of species. In particular, the whole text has been designed to promote or enhance the enthusiasm of the reader and hopefully inspire him or her to delve further into the depths of this fascinating subject. I have endeavored to include all information that will enable the beginner to learn the basics about reptiles, set up a terrarium, obtain animals, keep them successfully, and, hopefully, breed them.

John Coborn
Nanango, Queensland

Tortoises seem to keep growing in popularity and some can be very attractive. Photo by Susan C. Miller.

Facts about Reptiles

REPTILES AND MAN

Reptiles were creeping upon the earth's surface for millions of years before the human race evolved. Indeed, when the primitive ancestors of man appeared some 2.5 million years ago, reptiles had already reached a stage of evolution similar to that found in the contemporary species and the giant dinosaurs of earlier eras had been extinct for almost a hundred million years.

Primitive man had little knowledge of science; his prime purpose in life was to survive and to reproduce. His fellow animals were probably regarded as items of food or as dangerous adversaries. The facts that many snakes are venomous or large and dangerous (pythons) and that crocodiles are fiercely predatory probably has a bearing on the many attitudes, myths, and legends pertaining to them. Our knowledge of the natural sciences has increased many thousand fold in recent years, but some of these primitive attitudes still persist in many sections of the community. In the average household, the very mention of the word "snake" will produce shudders and grimaces. We hear such statements as "I hate snakes" and "the only good snake is a dead one" all the time, but do the people who utter these statements really understand any snake enough to be able to judge it in such a way? In most cases, the attitude is derived only from hearsay or from what may have been seen in popular movies. In any case, the vast majority of the earth's town and city dwellers are never likely to see a snake in real life other than in a zoo or pet shop.

The apparently universal loathing of one of God's creatures is perhaps easy to understand. The serpent's evil reputation began in the Garden of Eden, and it seemed that the Creator invented it solely to be nasty to people, but we have since discovered that He also had other motives! Although only about 10% of the world's 3,000 or so species of snakes are venomous, and an even smaller percentage dangerously so, the serpent has almost always been a symbol of evil. And this applies to the 90% of snakes that are harmless as much as to the dangerously venomous ones! Indeed, the harmless snakes (and even some creatures that are not snakes at all) are made to pay, often with their lives, for their relationship to or even vague similarity to venomous species.

Snakes make really good subjects for sensationalism. They are ideal creepy material in thriller movies, and even if a venomous species is supposed to be represented, usually only harmless species are used on the set for reasons of safety as "most people won't know the difference anyway," an unfortunate disregard for correctness in the education of the masses. Geographical correctness is also disregarded in such cases, and a South American boa constrictor can just as easily turn up on safari in Africa or a North American king snake can appear in some hotel bedroom on the French Riviera!

The fear of the crocodile is perhaps easier to understand. The simple fact is, if you jump into a river containing large crocodiles, you are likely to be eaten. However, those who are sensible enough to avoid such situations are never likely to get into such danger.

Tortoises, turtles, and terrapins are perhaps the least feared group of reptiles. In fact, many of those with an ignorant attitude toward reptiles in general will look at you in amazement if you tell them that a tortoise is more closely related to a snake than it is to a guinea pig! Tortoises once were the "standard pets" of the reptile world, though this has

Turtles are perhaps the least feared of all reptiles. Shown here is a baby Diamond-back Terrapin, *Malaclemys terrapin rhizophorarum*, photographed by William B. Allen, Jr.

FACTS ABOUT REPTILES

The interest in skinks by the herp hobby has always been somewhat minor, although fairly steady. Photo of *Eumeces laticeps* by Isabelle Francais.

diminished somewhat in recent years due to over-collecting, poor husbandry, and the subsequent protection of remaining wild populations.

Attitudes toward lizards are mixed. The person with a fear of snakes will almost certainly have some revulsion toward lizards. In some countries certain lizards are regarded as being highly venomous. The Leopard Gecko of southern Asia, for example, is greatly feared by the natives, while the good old Blue-tongued Skink of Australia is frequently regarded as a "Puff Adder" or "Death Adder" by many country people. In the USA, the Red-headed Skink also has a dubious reputation in many areas. One understandable reason for fearing lizards is that some of them are limbless or have very short limbs, giving them a very serpentine appearance. Of course, there are a couple of lizard species (genus *Heloderma*) that are venomous, and these are, understandably, feared in the areas where they occur.

CONSERVATION

Being unable to adapt to new environmental

conditions as well as many mammals and birds, reptiles have suffered much in terms of numbers as man changes the landscapes to suit himself. Older readers with an interest in nature will remember that lizards and snakes were much more frequently seen a few decades ago than they are today. Forty years ago the numbers of turtles, lizards, and snakes to be found seemed to have been almost infinite. But today the majority of these have disappeared. To even have a

This is the rare and attractive Bog Turtle, *Clemmys muhlenbergii*. Once common, its habitat has been greatly reduced by man over the last three decades, and it now holds a place on the U.S. Federal Endangered List. Photo by R. T. Zappalorti.

The Northern Pine Snake, *Pituophis melanoleucus melanoleucus*, also is becoming quite rare and is protected in many parts of its range. Photo by R. T. Zappalorti.

FACTS ABOUT REPTILES

remote chance of seeing a wild reptile today most of us have to travel some considerable distance into more remote country. Populations of reptiles in some locations have decreased in numbers by 50% to 100%. In other words, certain species in Europe, North America, and other industrialized areas have become extinct in portions of their range or at the best become exceedingly rare.

Only recently have herpetologists begun to learn more about the very rare blind salamanders. This one is the Georgia Blind Salamander, *Haideotriton wallacei*. Photo by R. D. Bartlett.

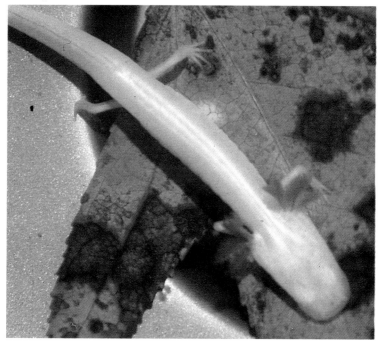

What then has contributed to this decline in numbers? There are a number of factors which can be considered.

* Habitat Destruction: During the fifties through to the seventies, reclamation of wetlands became something of a fashion. Reclaimed land could be used for forestry, agriculture, and residential or industrial building. There was no thought for the reptiles (indeed, reptiles would no doubt be the last, or almost the last, consideration by developers) or other creatures that depend on wetlands to be able to survive. Fortunately the last two decades have seen widespread recognition that wetlands are ecologically important; however, in some areas this is already too late. Destruction of forests, heathlands, and other habitats for similar reasons to those above has contributed to the decline of many species. What is required is a general international strategy to control the environment and to maintain its ecology intact by balanced use of the land.

* Pollution: Chemical effluents released from factories or as by-products of mining, etc., have had a dramatic destructive effect on aquatic life at various times and locations. The problem has been more or less alleviated in countries that have instituted strict legislation on the disposal of industrial effluents. However, in most parts of the world this kind of pollution must still be regarded as a serious threat to fish and amphibian populations, thus indirectly affecting the many reptiles that feed on them. Acid rain caused by the drift of industrial smoke containing sulfurous gases is another factor contributing to the demise of aquatic invertebrate, fish, and amphibian populations and

ultimately the reptiles and higher vertebrates in the food chain. An even more serious threat is probably the use of biocides and inorganic fertilizers in agriculture. Biocides include herbicides (weed killers) and insecticides that are in some cases so potent as to kill small reptiles directly. They are also indirectly affected by a severe loss of prey animals or by the destruction of vegetative cover. Fortunately, western

One of the rarest snakes in the U.S., the San Francisco Garter Snake, *Thamnophis sirtalis tetrataenia*, is heavily protected. Photo by C. Banks.

societies at least are trying to address the problem by developing safer chemicals.

*Collecting and Trade: The collecting and trading of reptiles for private collections, for research purposes, and for skins and food has, in the past, been a major factor contributing to the demise of wild populations. An anonymous article entitled "The exploitation of reptiles and fish for trade purposes," published in a British scientific journal in 1928, shows that concern was already being expressed at that time. Among other animals, the article reports "tens of thousands of Moorish tortoises, *Testudo graeca*, tens of thousands of green tree frogs, *Hyla arborea* (one single consignment amounted to 10,000 specimens), hundreds of cave salamanders, *Hydromantes genei* and spectacled salamanders, *Salamandrina terdigitata*" were imported into England. More recently it was estimated that between 1968 and 1970, 66,930 turtles (including land tortoises and marine turtles), 102,890 lizards, and 65,970 snakes were

One of the most popular animals in the pet trade is the Green Anole, *Anolis carolinensis*. Photo by Elaine Redford.

FACTS ABOUT REPTILES

Once available in department stores, juvenile Red-eared Sliders, *Pseudemys scripta elegans* are now only available on a restricted basis.

collected in Italy for commercial, scientific, and culinary purposes! During 1971, Yugoslavia exported a massive 400,000 Hermann's tortoises to various western European countries; it is estimated that that the survival rate of these tortoises in alien climates was less than 20% in the first year, so you can appreciate what the annual market was for these animals! These are mere examples of the potential destructive power of over-collection. Fortunately, legislation in most countries has now regulated the trade of many species thought to be endangered.

*Introduced Species: Introduction of non-native animal species to new areas can have a negative effect on

FACTS ABOUT REPTILES

If someone were to introduce a foreign species into this Jordan's Salamander, *Plethodon jordani*, environment, the salamander could very easily become prey and die out. Photo by R. T. Zappalorti.

the native reptiles. Thus fish introduced for economic purposes will soon destroy an amphibian breeding habitat and indirectly affect reptiles that prey on amphibians. Even domestic cats, dogs, and other feral animals will take their toll. The mongoose, introduced to some areas to control snakes, is also a major predator of lizards and amphibians. The pheasant, introduced into Luxembourg, has been shown to have contributed to the demise of the slow worm, *Anguis fragilis*. Another example is the introduction of the marine toad, *Bufo marinus*, into Queensland, Australia (where it is known as the "cane toad"), in the 1930s, ostensibly to control pests of the sugar cane. The toads found the climate and conditions so much to their liking that they have been on the increase ever since and are destroying many of the smaller native animals, including lizards and small snakes, both by directly devouring them and by displacing populations. Ironically, they have never been really adept in the job for which they were first introduced.

EVOLUTION OF THE REPTILES

To better understand the position of modern reptiles in zoological classification, it helps to know a little about their evolution. All forms of life probably originated in water, and the first life-forms to pioneer a life on land were certain plants and invertebrates. The first vertebrate (back-boned) animals were fishes, which have an evolution story all of their own, but it will suffice to say here that about 380 million years ago, during what geologists call the Devonian period, the first fishes started to creep out of the water and onto the land. Known as lobe-finned fishes (due to the tough, limb-like character of the fins), they led directly to the first amphibians, which developed over a period of 35 million years.

Many species of amphibia developed and were perfected during the Carboniferous period that followed the Devonian. Some of these became more terrestrial than others so that they could exploit the invertebrate prey that already lived on the land. After a time, many only needed to return to water for the purpose of reproducing, though they always required reasonably high humidity in order to avoid desiccation.

In order to exploit terrestrial life even further, it was thus necessary to evolve means of conserving moisture within the body. The answer arose in an extra protective layer over the skin, which eventually developed into the typical reptilian scales. Reproduction remained a problem, however, and a means of preventing desiccation in eggs laid on land was necessary. Unlike the amphibians, which had to return to water in order to lay their eggs (as typical modern amphibians still do), the emerging reptiles found a means of reproducing on land, often some considerable distance from water. This phenomenon became possible due to the development of internal fertilization and an egg with a thick, protective, water-conserving shell. These eggs could be concealed away from predators on the dry land, buried below the surface, or hidden among dense vegetation. These first eggs were large-yolked and the embryos could develop to a reasonably advanced state inside the egg, thus avoiding the vulnerable larval stage of the

Reptile eggs have evolved into the tough, leathery containers they are now as an adaptation to terrestrial life. Photo of the Eastern Hognose Snake, *Heterodon platirhinos*, by William B. Allen, Jr.

amphibians. Enclosed in a sort of sac called the amniotic membrane, the embryo grew in the water-like contents, receiving oxygen through the membrane walls and disposing of carbon dioxide by means of the allantois, a further membrane. Another membrane, the chorion, surrounded the whole contents just inside the tough, leathery outer shell. (This sequence of development is of course theoretical, based on what is known about the eggs of more primitive living reptiles and what must have happened to change an amphibian into a reptile, for early reptile eggs preserved as fossils certainly do not show soft parts such as the yolk and membranes.)

The reptiles became so successful on the land that they soon (in geological terms) ruled the earth. They developed diverse forms and evermore efficient ways of preying on invertebrates or on each other. No herbivorous reptiles had yet developed during the early Permian period, so it was a tremendously competitive time, as various adaptations for gripping, tearing, and chewing, as well as means of protection, had to be developed. Limbs lengthened and moved under the body rather than being at the sides, thus allowing greater speed and stamina. Herbivores developed to take advantage of the prolific plant life. Para-mammals and true mammals began to evolve at this period, as did the dinosaurs, which kept all other forms of life suppressed for the next 140 million years, right through to the end of the Cretaceous.

The main ancestors of the crocodilians developed during this period, and these reptiles have changed little through to modern times. The rhynchocephalians also developed at this time, but the only surviving member of this group today is the Tuatara, *Sphenodon punctatus*, of New Zealand. The Chelonia probably started as a separate group around the late Permian and were at their most prolific during the Jurassic and Cretaceous periods. In several respects they can be classed as the most primitive of the living reptiles.

Our modern snakes and lizards probably arose from groups of Rhynchosauria, the same ancestors of the Tuatara, but split from the line during the early Jurassic and, indeed, split into their two respective suborders toward the end of

FACTS ABOUT REPTILES

Facing Page: In a "class" by itself, the Tuatara, *Sphenodon punctatus*, is the one and only living member of its entire order, Rhynchocephalia. Photo by R. T. Zappalorti. Above: The "latest" herpetiles to evolve were the snakes. Pictured here is the breathtaking Sonoran Mountain Kingsnake, *Lampropeltis pyromelana pyromelana*. Photo by Louis Porras.

the same period. The oldest ancestors of our modern lizards are seen in fossils from the late Jurassic, about 200 million years old; interestingly, the first fossil bird ancestors appear at about the same time. Snakes pose a great evolutionary problem. Very little is known about their early history, though they probably arose from the lizard line during the Cretaceous. Due to the brittleness of the bones of small lizards and snakes, fossil study material is extremely sparse, though it can be theorized that snakes arose from burrowing lizard stock. However, for the time being, a large part of snake evolution remains a mystery.

Classification of the Reptiles

The modern system of zoological classification was pioneered by Karl von Linne (Linnaeus) (1707-1778), a Swedish biologist (actually more of a botanist than a zoologist, but his system works for animals as well as plants) who saw the need for a more logical listing of the thousands of existing known kinds of animals and plants plus the numerous new ones that were being discovered at frequent intervals as man's explorations of remote areas of the Earth intensified. Linnaeus took upon himself the prodigious task of naming every one of these different living things. To do this he created his system of binomial nomenclature, the practice of giving each species a double Latin name.

At that time the science of zoology was still in its infancy and internationally inconsistent. Although many animals had common names, these names differed from language to language and even from dialect to dialect. In addition, animals had not been studied in depth, and the different groups were not adequately distinguished from each other. For example, all amphibians were included with the reptiles; a lizard was a lizard, but a newt was a water lizard. We now know just how different these groups really are.

In Linnaeus's system of taxonomy (taxonomy being the science of naming

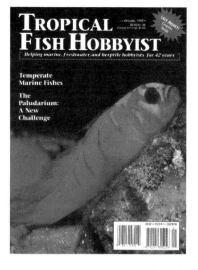

Since 1952, *Tropical Fish Hobbyist* has been the source of accurate, up-to-the-minute, and fascinating information on every facet of the aquarium hobby. Join the more than 50,000 devoted readers worldwide who wouldn't miss a single issue.

Subscribe right now so you don't miss a single copy!

Return To:
Tropical Fish Hobbyist, P.O. Box 427, Neptune, NJ 07753-0427

YES! Please enter my subscription to *Tropical Fish Hobbyist*. Payment for the length I've selected is enclosed. U.S. funds only.

CHECK ONE:
- ❏ 1 year-$30 / 12 ISSUES
- ❏ 2 years-$55 / 24 ISSUES
- ❏ 3 years-$75 / 36 ISSUES
- ❏ 5 years-$120 / 60 ISSUES

(Please allow 4-6 weeks for your subscription to start.) *Prices subject to change without notice*

- ❏ LIFETIME SUBSCRIPTION (max 30 Years) $495
- ❏ SAMPLE ISSUE $3.50
- ❏ GIFT SUBSCRIPTION. Please send a card announcing this gift. I would like the card to read: _____
- ❏ I don't want to subscribe right now, but I'd like to have one of your FREE catalogs listing books about pets. Please send catalog to:

SHIP TO:
Name _____
Street _____ Apt. No. _____
City _____ State _____ Zip _____

U.S. Funds Only. Canada add $11.00 per year; Foreign add $16.00 per year.
Charge my: ❏ VISA ❏ MASTER CHARGE ❏ PAYMENT ENCLOSED

Card Number _____ Expiration Date _____
Cardholder's Name (if different from "Ship to":) _____
Cardholder's Address (if different from "Ship to":) _____
Cardholder's Signature _____

...From T.F.H., the world's largest publisher of bird books, a new bird magazine for birdkeepers all over the world...

CAGED BIRD HOBBYIST
IS FOR EVERYONE
WHO LOVES BIRDS.

CAGED BIRD HOBBYIST
IS PACKED WITH VALUABLE
INFORMATION SHOWING HOW
TO FEED, HOUSE, TRAIN AND CARE
FOR ALL TYPES OF BIRDS.

Subscribe right now so you don't miss a single copy!   SM-316

Return to:
CAGED BIRD HOBBYIST, P.O. Box 427, Neptune, NJ 07753-0427

YES! Please enter my subscription to **CAGED BIRD HOBBYIST**. Payment for the number of issues I've selected is enclosed. *U.S. funds only.

CHECK ONE:		
☐	4 Issues	$9.00
☐	12 Issues for the Price of 10	25.00
☐	1 Sample Issue	3.00

☐ Gift Subscription. Please send a card announcing this gift.   PRICES SUBJECT TO CHANGE
I would like the card to read _____

☐ I don't want to subscribe right now, but, I'd like to receive one of your FREE catalogs listing books about pets. Please send the catalog to:

SHIP TO:
Name _____ Phone (   )
Street _____
City _____ State _____ Zip _____

U.S. Funds Only. Canada, add $1.00 per issue; Foreign, add $1.50 per issue.

Charge my: ☐ VISA   ☐ MASTER CHARGE   ☐ PAYMENT ENCLOSED

Card Number _____ Expiration Date _____

Cardholder's Name (if different from "Ship to:") _____

Cardholder's Address (if different from "Ship to:") _____

Please allow 4–6 weeks for your subscription to start.   Cardholder's Signature

CLASSIFICATION OF REPTILES

animals and plants), the first name is the generic name, the second is the specific name. The generic name is the name of the genus, a group of closely related species, while the specific name identifies the species within the genus. For example, the lizard genus *Lacerta* has a number of European, Asian, and African species. *Lacerta vivipara* is the European Common or Viviparous lizard, *Lacerta viridis* the Green Lizard, and *Lacerta agilis* the Sand Lizard. *Lacerta* is, in fact, the Latin word for lizard, and Linnaeus probably used this as these local types could well have been the first lizards he gave his attention to. The respective meanings of each of the specific names of the three examples given are "viviparous," "green," and "agile." In the first two examples it can be seen that the Latin name follows the English name fairly closely, but the third example has

This little fellow's Latin name is *Pseudemys scripta elegans*. *Elegans* means "elegant," and *scripta* means "written." Photo by Michael Gilroy.

CLASSIFICATION OF REPTILES

Latin and English names with different meanings.

As the Latin names are international, a zoologist will understand what species is being referred to, whatever his own native language is. The following table will illustrate just how useful this internationalism is:

Latin	*Lacerta vivipara*	*Lacerta viridis*	*Lacerta agilis*
English	Common Lizard	Green Lizard	Sand Lizard
French	Lezard Vivipare (Viviparous Lizard)	Lezard Vert (Green Lizard)	Lezard des (Wood Lizard)
German	Waldeidechse (Wood Lizard)	Smaragdeidechse (Emerald Lizard)	Zauneidechse (Fence Lizard)

In Latin terms, this Collared Lizard is known as *Crotaphytus collaris*. In this case, the second, or species, name is the one its common name is based on. Photo by Elaine Radford.

CLASSIFICATION OF REPTILES 35

Without a doubt, one of the most misleading Latin monikers is the Racer's *Coluber constrictor*, since it is not a constrictor at all. Photo by R. T. Zappalorti.

It may be noted that there is no conformity in the common names of these examples (English equivalents of names have been given in parentheses below those in French and German).

By today's standards, the new system was somewhat primitive, but it was at least a start. Once a number of species had been named, the next task was to arrange them in some kind of logical order. Natural classification is a hierarchical arrangement of animals (or plants) into different groups, based on differences and similarities between them. In this arrangement, the lowest rank is the species, which can be described as one of a group of organisms that are all essentially the same, at least with minimal variation, and that interbreed to produce more individuals of the same type.

CLASSIFICATION OF REPTILES

When a number of species are very similar to each other, having many aspects in common but not quite the same, they are placed together in a genus. The species used as examples above are all members of the genus *Lacerta* because they all have certain similarities.

Numbers of similar genera (plural of genus) are placed together in a family; similar families are placed into an order, orders into classes, and so on.

The following table illustrates how the four orders of reptiles fit into the general zoological picture:

KINGDOM	Animalia			All Animals
PHYLUM	Chordata			Animals with notochord
SUBPHYLUM	Vertebrata			Vertebrates
CLASS	Reptilia			All Reptiles
ORDER	Chelonia (turtles)	Crocodilia (crocodiles)	Rhynchocephalia (tuatara)	Squamata (lizards, snakes, amphisbaenians)

It can be seen here that the lizards and snakes share an order together. This is because they have many similarities, so they are separated into suborders (along with the amphisbaenians). A typical classification of a snake, an amphisbaenian, and a lizard species would be as follows:

ORDER	Squamata (lizards, snakes, and amphisbaenians)		
SUBORDER	Lacertilia (lizards)	Amphisbaenia (amphisbaenians)	Serpentes (snakes)
FAMILY	Lacertidae	Amphisbaenidae	Colubridae
GENUS	*Lacerta*	*Rhineura*	*Elaphe*
SPECIES	*L. viridis* (Green Lizard)	*R. floridana* (Worm Lizard)	*E. obsoleta* (Rat Snake)

CLASSIFICATION OF REPTILES

The orders and suborders are perhaps the most important categories when regarding the class Reptilia as a whole.

As our zoological knowledge has increased, the science of taxonomy has been gradually improved. Nowadays, there are strict regulations regarding the naming and publishing of newly discovered animal species, and these are governed by the *International Code of Zoological Nomenclature*.

The provision of a clean-cut binomial to a species is not without problems, however, and in some cases it is necessary to add a third

The attractive Glossy Snakes are of the genus *Arizona*, which means "dry land". Photo by B. Kahl.

CLASSIFICATION OF REPTILES

name, making a trinomial. This is used when geographical groups of certain species show differences, but are not sufficiently different to be considered as separate species. Such groups of animals are known as subspecies. While many reptiles are regarded just as species, with a binomial, some may have quite a large number of subspecies. When a species is relegated to subspecific rank, the first or original species described has its specific name simply repeated, while further subspecies are assigned a different subspecific name. There are many examples, but to illustrate the point let us take a look at a genus of North American turtle species.

The Alabama Red-bellied Turtle of southern Alabama, USA, is considered to be a single species and thus has the simple binomial *Pseudemys alabamensis*. Conversely, the River Cooter in the same genus is considered to have two subspecies: *Pseudemys concinna concinna* and *Pseudemys concinna texana*. Other species may have many more subspecies. In the same genus for example, the Slider, *Pseudemys scripta*, 14 subspecies have been described.

The binomial and trinomial names are usually written in italic or underlined script to avoid confusion with common names or other parts of the text in which they are cited. When it is necessary to use the binomials or trinomials several times in a given text, abbreviations are accepted, provided they have been written once in full. Thus *Pseudemys alabamensis* can be reduced to *P. alabamensis*, and *Pseudemys concinna concinna* may be abbreviated to *P. c. concinna*.

CLASSIFICATION OF REPTILES

The "binomial" name for this Loggerhead Musk Turtle is *Sternotherus minor*. "Minor", the suffix, means small or secondary. Photo by K. T. Nemuras.

Housing

There are various types of housing used for captive reptiles, ranging from small indoor tanks to large outdoor enclosures. There are no hard and fast rules regarding types of accommodation, and it often reflects the taste of the individual hobbyist. However, there are a few ground rules to be considered before any specimens are obtained. An indoor cage in which reptiles are kept may be called a vivarium or terrarium (the author prefers the latter). Such a terrarium should obviously be escape-proof, must be of adequate dimensions for the animals to be housed, and must be equipped with the essential life-support systems needed by the species being kept (heating, lighting, humidity, ventilation, etc.). It should be easy to clean and maintain and should be as attractive in appearance as possible.

TYPES OF TERRARIA

There are a number of categories into which terraria can be placed, depending on the kind of habitat to be reproduced. The main types are as follows:

1) The heated dry terrarium—for desert and semi-desert species.

2) The heated humid terrarium—for tropical rainforest species.

3) The unheated dry terrarium—for species native to temperate heathland or prairie, etc.

4) The unheated humid terrarium—for species of the damp temperate woodlands.

5) The aqua-terrarium—for species that require a large body of water. This may be tropical or

HOUSING

temperate, depending on the species being kept.

The shape of the individual terrarium is unimportant, providing the points mentioned above are taken into consideration. However, it does make sense to use fairly tall terraria for arboreal species and relatively shallow ones for terrestrial types. Commercial terraria, complete with all their life-support systems, are now generally available through pet shops that cater to herps (herps is a shorthand way of referring to reptiles and amphibians, a rather crudely formed but effective and easy to remember word), but many enthusiasts still prefer to build their own. All manner of materials may be used in terrarium construction, and it is often interesting (and economical) to experiment with various items you may obtain from second-hand or do-it-yourself stores, or even from the local dump; old TV or hi-fi cabinets often have quite attractive possibilities!

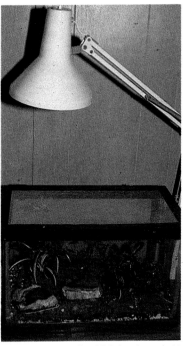

Here is an example of lighting that would be adequate for most non-breeding snakes, but not "nutritional" enough for turtles or lizards. Photo by Elaine Radford.

TERRARIUM SIZES

Dimensions of terraria will, of course, be dictated by the number, size, and

HOUSING

These beautifully constructed plastic "laid-back" tanks have become enormously popular throughout the hobby over the last decade. Photo by William B. Allen, Jr.

habits of the creatures you intend to keep. Most reptiles have a very small range of territory in the wild. If food and water were available close to its shelter at all times, the average snake would be quite content to spend its whole life in an area of one or two square meters (1-2 sq yds). Again, there are no hard and fast rules, and usually all you require is a little common sense to decide on the dimensions of your terrarium. For example, a pair of 20 cm (8 in) terrestrial lizards could be kept comfortably in a terrarium 60 cm long x 30 cm deep x 30 cm high (2 x 1 x 1 ft). For similar sized arboreal lizards the same dimensions can be used but the terrarium is arranged upright, the longest length being the height. The same

HOUSING

cages could be used for terrestrial or arboreal snakes up to 45 cm (18 in) in length. On the other end of the scale, a 3 m (10 ft) python or boa would require a cage about 2 m long x 1 m deep x 1 m high (approx. 6 x 3 x 3 ft), though such snakes are often kept quite successfully in smaller cages.

WOODEN TERRARIA

Various types of wood can be used to construct a low humidity terrarium, but timber is really not suitable for use in permanently damp situations as even with a few coats of varnish it will quickly deteriorate. A simple box-type terrarium with a framed glass front is easy to make using 12 mm (½ in) plywood. After carefully cutting the panels to size (or having an expert cut them for you), the top, bottom, and ends are simply glued and tacked together. The glass viewing panel may be simply slid into grooves on the top and bottom or in the sides. Alternatively you may wish to have a more professional finish by mounting the glass in a wooden frame, then hinging this as a door onto the body of the terrarium.

Ventilation holes should be drilled in groups in the ends of the cage about a third the distance up from the base to the top, and further holes should be drilled in the top of the terrarium to encourage ventilation by convection (hot air rises through the top, pulling replacement fresh air through the lower sides). Instead of individual holes, a more attractive finish can be attained by cutting out large square panels and covering them with fine mesh (insect screening is fine for small reptiles, but galvanized weld-wire mesh should be used for more powerful species) held in place by a frame of narrow wooden molding. A sliding metal,

HOUSING

plastic, or wooden tray can be fitted in the base of the box to hold the substrate and facilitate cleaning.

GLASS TERRARIA

In the past, simple aquarium tanks commonly were used to keep many species. These are indeed quite suitable to use providing you can provide adequate ventilation. With modern silicone rubber adhesives it is possible to build glass terraria in many shapes and sizes, and by using a combination of glass

HOUSING

and acrylic (plexiglass) materials, you can still drill ventilation holes in the ends or the back. Glass is extremely good for making humid terraria or aqua-terraria in which part of the floor is to be a permanent water feature. The lid for a glass terrarium is best fashioned from plywood or plastic and constructed to form a cavity in which the heating and lighting apparatus can be concealed from view from the outside. Such an apparatus should be separated from the main body of the terrarium by a wire mesh screen so that the animals cannot burn or even electrocute themselves.

THE BUILT-IN TERRARIUM

A permanent, fixed terrarium built into part of the house or apartment can be extremely attractive and will perhaps make a focal point in the lounge, family room, conservatory, or wherever. Such a structure can be built into an alcove or be free-standing. Really substantial terraria, especially those suitable for large species such as green iguanas, monitor lizards, pythons, large turtles, or even crocodiles, can be constructed from bricks or concrete blocks and can

If you have some of the larger herps (i.e., boas, pythons, water dragons, etc.), your dealer will be able to supply you with a tank of adequate size. If not, then perhaps building your own would be ideal. Photo by David Alderton.

include a permanent, drainable, concrete pool with a filtration system, waterfall, and so on. Artificial cliff-faces with ledges and controllable hiding caves may be constructed as strong, permanent features.

For smaller reptiles, you can include pockets among the rockwork in which you can arrange potted plants, but larger reptiles would soon destroy these with their weight, so a compromise (perhaps more decorative rockwork and logs, etc., or even plastic plants) may be necessary. A visit to the reptile house in your local zoo (or as many zoos as you can if you get the chance) will no doubt give you some excellent ideas for inclusion in your home terrarium.

A word of warning: Before commencing any major construction, be sure you are not contravening any regulations. It may be a good idea to get engineering or plumbing advice before introducing weighty structures into your home. You would not, for example, build a concrete structure on a suspended wooden

Who says herpetology can't be attractive? Even the most undistinguished animals would look good in this gorgeous domestic setup. Photo by Susan C. Miller and Hugh Miller.

HOUSING

floor! A concrete floor at ground level is the best bet.

OUTDOOR ENCLOSURES

If you are interested in reptiles that hail from a climate similar to your own, you may consider keeping some in a walled outdoor enclosure. Even some species from warmer climates than your own may

be placed outside during the warmer months of temperate summers (but watch out for sudden weather changes) and they will gain enormous benefit from the fresh air and sunlight. Captive reptiles will be able to live almost natural lives in an outdoor enclosure. Insectivorous species will be able to forage for much of their own food. If you have a sizable pond in the enclosure, you can breed frogs or small fish that will form part of the diet of many snake species. Additionally, reptiles in such conditions will be relatively more likely to behave territorially, providing them with a greater stimulus to breed than those kept in smaller indoor terraria. However, any eggs laid by exotic species may have to be removed for artificial incubation.

A walled outdoor enclosure usually consists of a central mound surrounded by a continuous wall. The area enclosed will depend on the amount of available space, what you can afford, and the size, species, and numbers of reptiles you intend to keep. The larger the enclosure, the more difficulty you will have in observing the inmates. The area must be one that receives a full quota of sunlight, and you should arrange to have several open basking sites. The wall foundations should be at least 50 cm (20 in) deep to prevent inmates from escaping by burrowing. The earth removed for the foundations can be used as a base for the central mound, which can be further enlarged and landscaped by using rocks and topsoil.

For most smaller snakes and lizards the height of the wall need not be more than 90 cm (3 ft), but for snakes over 90 cm (3 ft) in length, you will have to dig a deeper trench so that the inside of the wall is higher than the

HOUSING

outside. The height on the inside should be at least 20 cm (8 in) more than the length of the longest snake. Also beware that no part of the central mound is near enough to the wall to allow escapes. You must also watch for plant growth near or on the wall that could provide an escape route.

The wall may be built of concrete blocks or bricks and, on the inside at least, should be smoothly rendered with a fine cement mortar. To prevent the more agile lizards from escaping, an overhang at least 15 cm wide should be affixed at the top of the wall. Alternatively, a continuous strip 15 cm (6 in) wide of smooth plastic laminate can be affixed to the inside of the wall just below the top. A continuous row of glossy ceramic tiles will do the same job. This will stop most lizards from getting a grip with their claws, but needless to say, it will not stop geckos or some anoles.

It is best to leave a space about 90 cm (3 ft) wide between the wall and the start of the central mound. This can be in the form of a concrete moat containing water, or it can be covered with clean gravel that is kept free of plants and debris; this will discourage animals from spending time there and devising means of getting over the wall! Hibernacula (which may also be used as sleeping quarters) for native species can be included while you are constructing the central mound. Plastic food boxes or similar items can be used. A hole is made in one side of the box and a plastic (5 sq cm in diameter) drain pipe can be fitted through the hole. The box should be packed loosely with hay or dry sphagnum moss and buried at least 60 cm (2 ft) deep in the mound, with the pipe sloping downward to a side of the mound (to prevent rain entering). The entrance to the pipe can be

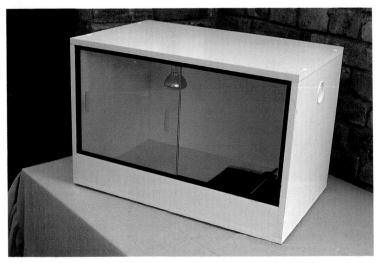

Shown earlier, this attractive little tank can be easily made with some time, patience, and a bit of carpentry skill. Photo by Susan C. Miller and Hugh Miller.

disguised with rocks, leaving just a fissure through which the reptiles can pass. If you want greater control over your reptiles, controllable hiding places can be included by using concrete slabs, etc.

If you include a pond, waterfall, and stream in your reptile habitat it can become a very attractive feature of the garden. The mound can be tastefully decorated with rocks and logs and planted with various rock and alpine plants and a few shrubs. Flowering plants will attract flying insects, which will be a welcome addition to the menu, though it will still be necessary to supplement the diet of insectivorous species on a regular basis. Much of the mound should face toward the prevailing direction of the sun (to the

HOUSING

south, in the Northern Hemisphere) so that plenty of basking areas are available. It is wise to provide a few plant-free sandy areas where breeding reptiles can burrow and lay their eggs. With native temperate species or those from a similar climate, the eggs can be left to hatch naturally, but those of sub-tropical and tropical species must be removed for artificial incubation.

GREENHOUSES

The nearest compromise to a year-round outdoor enclosure for sub-tropical and tropical reptiles in a temperate climate is a heated greenhouse. The greatest disadvantage is probably the expense of creating and maintaining the right conditions in such a large volume, but if you can overcome this you will not regret it. A miniature rain forest may be created and high levels of humidity maintained by using sprinklers, heated pools, and waterfalls. Heat lamps should be placed in several positions, giving the reptiles a choice of basking sites (while making sure that the lamps are well-protected from water). The background temperature should not fall below 15°C (59°F) at night and care should be taken to ensure that, during hot weather, overheating does not occur. To prevent escapes, windows and ventilation holes should be covered with double mesh screens. Heating apparatus such as pipes, radiators, and lamps should be boxed or caged-in to prevent the reptiles from burning themselves.

INTERIOR DESIGN FOR TERRARIA

Indoor terraria require various furnishings that are both functional and decorative. Although it is possible to keep and even breed many reptiles in almost "clinical" conditions (a sheet of absorbent paper, a water dish, and a hide box as the only decorations), most herpetologists require their terraria to be decorative and to at least simulate natural conditions. However, the clinical method still has its uses and may be convenient if you are breeding and rearing reptiles on a semi-professional basis.

Substrate Materials:
There is a remarkable range of materials suitable for use as floor coverings in the terrarium. Coarse sand can be used for the desert terrarium, while a mixture of peat, sand, and loam is useful in woodland terraria. For larger, robust species it is best to use washed gravel, which can be obtained in various grades. Pea-sized gravel is useful for small to medium reptiles, while 2-3 cm (1 in) shingle can be used for larger pythons, monitor lizards, and so on. The beauty of gravel, of course, is that it can be thoroughly washed, disinfected if necessary, and re-used. It is best to have a spare store of gravel so that you can replace that in the terrarium at any time necessary. The used gravel can then be cleaned, dried, and stored for future use. Do not use very fine sand, as it tends to cake between the scales of some reptiles, can cause stress, and can trigger other unpleasant skin diseases or conditions. For burrowing species, a mixture of coarse sand and peat or leaf litter can be used (house plant potting mixture is ideal). Whatever substrate material you use, it must be removed and washed or replaced at regular intervals, more frequently for those larger species with copious droppings.

Rocks:
Rocks may be found in many interesting shapes and colors. They are not

HOUSING

Although a tank like this may seem a little complicated to deal with, in the long run it would be ultimately worth it for your pets. Artwork by Scott Boldt.

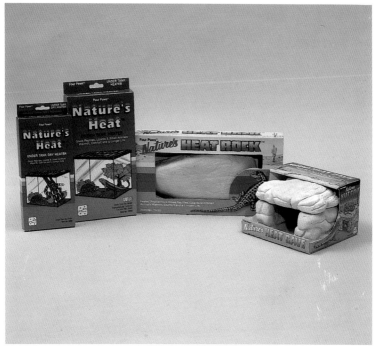

Four Paws offers a wide range of heat sources for your reptiles, including heated rocks and caves for the interior of your tank and under-tank heaters that warm the entire habitat.

only decorative in the terrarium, they are used as basking sites, as hiding places, and as an aid to sloughing. You may be able to buy suitable rocks at your local pet shop, but it is much more interesting to go out searching for your own (ensuring, of course, that you have permission to remove materials from certain areas and that you do not cause damage to the environment). One potential problem is that natural rocks you collect might have chemical contaminants that will dissolve in the water, while purchased rocks should be safe even for use in an aquarium. Always make sure the rocks are arranged firmly in the terrarium so that they are unlikely to fall down and injure the animals. If large piles of rocks are used, it is best to cement them together, leaving a few controllable hiding places, but do not cement or glue them to the terrarium structure itself; they must be removable for cleaning and general maintenance.

Another possibility is the use of artificial rocks. Very

Incandescent bulbs that serve as sources of both light and heat and that are tailored to daytime or nighttime use, such as these from Energy Savers, are available at pet shops.

natural-looking and durable rocks can be made that are only a fraction the weight of the real thing. Pieces of foam plastic cut or torn into shape are used as a core. A mortar of three parts sand and one part cement, mixed with water to form a sticky consistency, is applied in a layer about 1 cm (0.5 in) thick over the foam, which has first been given a thick coat of wood glue or a similar bonding material. Leave the surface rough and allow it to set for about 24 hours. Prepare another mortar mixture and divide it into two or three batches, each of which can be given a different shade of concrete coloring (preferably browns, grays, yellows, reds, etc., but not too dark). Apply a coat of wood glue over the set rock and add a further layer of colored mortar over it. The different colors are blended into each other using a brush, and the rock surface can be sculpted into natural shapes using a small trowel. Finally, by dabbing and stippling with a barely damp brush (which must be cleaned out in water and shaken dry at regular intervals), you can apply the finishing touches to your rock. Two such rocks, one concave on the

This is a simple, yet perfectly acceptable, terrarium setup. Artwork by John R. Quinn.

HOUSING

top, the other concave on the bottom, can be made to fit exactly on top of each other, leaving a cavity that provides an excellent hiding place for lizards or snakes. A small entrance hole is left for the animals. You have instant access to them by simply lifting the upper rock. After setting, the rocks should be soaked in water for about 48 hours, rinsed, and dried before use; this step is necessary to remove any possibly dangerous chemicals.

Logs and Branches:

Hollow logs have tremendous possibilities in the terrarium for many species, but be sure that the animals are always easily accessible. Deep hollows are preferably cut open with a saw, then placed back together so that you maintain easy access. For arboreal species, a few tree branches are indispensable. It is more practical to use a dead tree branch and grow a creeping house plant on it than to try and grow a tree or shrub in the terrarium. Try to select branches with interesting shapes. Driftwood collected from the sea shore or from river banks is often attractive as it will have been weathered by the action of sand, sun, and water. All timber collected must be thoroughly scrubbed, rinsed, and dried before being placed in the terrarium. A log or branch immersed in a strong solution of bleach for a couple of days and then thoroughly rinsed in fresh water will take on a weathered appearance when it dries out.

PLANTS IN THE TERRARIUM

Healthy, living plants in the terrarium will provide a subtle esthetic touch. However, unless the terrarium is very large, it is futile to try and grow plants together with big, robust reptiles like monitor lizards,

HOUSING

The River Tank System RT30 combines a variety of elements into a complete ecosystem. Pools of water at different levels are connected by rapids or waterfalls and fish move freely from pool to pool. Plants grow hydroponically from a hidden gravel pocket in the rear which also acts as a biological filter. Plenty of room is available for aquatic animals and small reptiles above the river area including a "lizard ledge" which can be heated to provide an optimal environment. Photo courtesy Finn Strong Designs.

iguanas, pythons, or large turtles, as they (the plants) will be continually uprooted or flattened. Likewise, herbivorous lizards or tortoises will eat most plants before they get a chance to become established. In such cases, however unesthetic it may seem, you will have to compromise with tough artificial plants (fortunately, some quite realistic types are now available) or do without plants at all. For small lizards and snakes, the terrarium may be attractively planted. Select plant species compatible with the kind of terrarium and, if you are really pedantic, use those that come from the same habitat as the reptiles. Numerous house plants are suitable for the terrarium, and it will be useful to study a good book on house-plant culture to aid in your selection.

It is best to leave the plants in their pots, which can be concealed behind rocks or in special cavities in logs or rockwork. They are thus easy to remove and replace should they get "sick." It is advisable to have spare potted plants in similar size pots so that they can be changed at, say, monthly intervals or at cleaning time. The plant that has suffered the rigors of terrarium life can then be given a period of convalescence in the greenhouse or on the window ledge until its partner requires similar treatment.

LIFE-SUPPORT SYSTEMS

It is necessary to provide indoor terraria with artificial life-support systems as a compromise for the lack of natural conditions. Such systems will usually include temperature control, lighting, humidity, and ventilation. The first three will vary in necessity from species to species, but ventilation is important in all terraria.

Temperature Control:

Being ectothermic, many reptiles, especially lizards and snakes, maintain their body heat at a preferred temperature by moving in and out of warm places; by basking directly in the sun or absorbing heat from sun-warmed soil, rocks, or other items. The act of doing this is termed thermoregulation . Natural sunlight through terrarium glass has a propensity to serious overheating, so artificial means of heating have to be employed.

Temperatures required by individual species are detailed in the species descriptions later in the text. It is useful to remember, however, that with the exception of those species inhabiting lowland equatorial areas (which require a small reduction of only 2-3°C or 4-5°F), most reptiles require a fair reduction in temperature at night. This can be simply accomplished in home terraria by switching the heat source off in the evening and allowing the terrarium to cool to room temperature during the night. Summer and winter room temperatures in most dwellings are suitable for most reptiles during the night. Next morning the heat is simply switched on again.

Tungsten (Incandescent) Lamps:

Ordinary household light bulbs with tungsten elements had long been used as a sole source of heating and lighting in the home terrarium until it was discovered that the quality of light emitted was not good enough for diurnal basking lizards. However, tungsten bulbs do still have their uses in our hobby. They are inexpensive, emit a fair amount of warmth, and will supply supplementary light. The size of the terrarium will dictate the amount of wattage required to reach the required temperature. By experimenting with various wattages and a thermometer, a suitable temperature can be obtained. A tungsten bulb may be concealed inside a flowerpot or metal canister and controlled by a thermostat so that a constant minimum background temperature is maintained.

Spotlamps:

These are available in various types and may give off infrared or white radiant that which is very useful for directing at basking areas. Preferably, such lamps are set at one end of the terrarium only, thus creating a temperature gradation from one end of the cage to the other. The inmates will then be able to

HOUSING 61

Here you can see the usefulness of the spotlight heating technique. The animals can move away from it if and when they wish. Artwork by Scott Boldt.

select the areas in which the temperature is most comfortable to them. Ceramic plates or lamps are also available. These emit heat but no light and are useful for maintaining background heat at night.

Aquarium Heaters:
These are very useful for terraria requiring high

humidity. An aquarium heater placed in the pond will keep the water warm and the air warm and humid. The addition of an aquarium aerator will further increase humidity and will help maintain a suitable air temperature as well as provide additional ventilation and help keep the water fresh. An aquarium heater can even be used to heat the air space in a terrestrial terrarium by placing it in a concealed jar of water.

Cable Heaters and Pads:

These are used for heating parts of the substrate, and can be employed to provide further basking areas. Pads may be placed below the terrarium floor, while cables may be passed through the substrate itself.

Hot Rocks:

Several manufacturers sell a block of plaster or foam plastic containing a heating element and serving as an artificial rock on which a lizard or snake can bask. Such hot rocks are available in many different sizes and designs and are useful for desert species that require a very hot basking area. Since many lizards (and probably some snakes) regulate their basking time in response to the intensity of light from above and not based on the amount of heat below the body, accidents can happen unless basking is closely supervised.

Lighting:

Natural sunlight or a good substitute is essential to the well-being of all diurnal (day-active) reptiles, but especially lizards. It is probably the ultraviolet rays of sunlight that are most important as they help stimulate the manufacture

---

Facing Page: This drawing exemplifies simple yet effective lighting that almost anyone can provide with ease. Artwork by Richard Davis.

HOUSING 63

of vitamin D3 in the skin. This vitamin is essential for controlling the action of the important minerals calcium and phosphorus in the body, and without it various health problems will ensue. If possible, allow your reptiles the benefit of unfiltered natural sunlight as often as you can by allowing it to pass through mesh, rather than glass, into the terrarium. Small terraria can be moved out onto a balcony or into the garden in suitable weather conditions. Natural sunlight filtered through glass is not beneficial as the ultraviolet rays are cut out and there is the added danger of overheating. Be sure that there is some shade for the reptiles to use if they wish. In cool-temperate areas it is too cold to place tropical or sub-tropical terraria outside for most months of the year, so compromise lighting of good quality must be provided. Broad-spectrum fluorescent tubes which emit "blue" light will provide sufficient ultraviolet light for your lizards. Be aware that too much ultraviolet light can be more damaging than too little and that pure ultraviolet lamps should be used very sparingly indeed, if at all. Research into suitable light sources for horticulture, aquaria, and terraria is continuing. Information on suitable systems may be obtained from pet shops or directly from the manufacturers of lighting equipment.

Humidity:

Reptiles from moist regions will require a humid atmosphere, while others may require seasonal increases and decreases in relative humidity. The simplest (but most time-consuming) method of maintaining a suitably high humidity is to spray the interior of the terrarium several times a day with a fine mist of water. The maintenance of a high

HOUSING

humidity in the terrarium is also quite easy, however, if aquarium heaters are used in the water dish. Alternatively, the heater can be placed in a concealed jar of water that is topped up as necessary. As mentioned above, an aerator used in the water will further increase the humidity and air temperature and is ideal for use in tropical rainforest terraria. An attractive little waterfall or seepage can be created by using an airlift filter in conjunction with the aerator.

Poor ventilation in the terrarium will lead to a build up of stagnant, stale air and an excess of carbon dioxide, providing favorable conditions for disease organisms to multiply. In addition, an inadequately ventilated humid terrarium is likely to develop unattractive looking growths of molds and fungi on the substrate. It is therefore important to ensure that there is a constant air exchange in the terrarium, but without creating excessive cold drafts. In most cases, the provision of adequate ventilation holes in the sides and top of a terrarium is all that is required. The warmth generated by the heating apparatus will cause air convection currents, the warm air leaving through the top and fresh air replacing it through the side vents.

An aquarium aerator operated from a small air pump can be used to supply extra ventilation. If low humidity is required, the air outlet is not placed in water. If a terrarium is placed in a stuffy living room, particularly where tobacco smokers are present, it is recommended that the air inlet be placed outside the room, preferably in the fresh air. In cold conditions, the air tube can be passed near an underfloor heater or radiator so that the air is warmed before it is passed into the terrarium.

Safety Precautions:

As heating and lighting equipment is operated by electricity, you should take adequate safety precautions to avoid electrical accidents. Unless you are an adept electrician, use only equipment that has passed relevant safety standards and use it to the manufacturer's instructions. If in doubt, employ a qualified electrician to do your wiring and installation. Remember that electricity and water form a dangerous combination.

Another simple terrarium setup, this can be used for any reptile that you may come across, provided of course the tank is in the right proportion to the animal's size. Artwork by Scott Boldt.

General Management of Captive Reptiles

Captive animals are totally dependent on us, their keepers, for their welfare. General care includes such routines as feeding, cleaning, disease prevention, and so on. These routines are generally not too time consuming once you are organized, but they are important for the well-being of the animals in your care.

ACQUISITION OF SPECIMENS

There are various ways of acquiring specimens: directly from breeders, from pet shops, from specialist suppliers, or collection from the wild. Whatever means you have of acquiring specimens, ensure that you are not breaking any conservation laws, bearing in mind that many reptile species are protected by law in many countries. It is only by having a responsible attitude to our wildlife that we can hope to preserve it for future generations. Laws may vary from country to country or even from state to state or town to town, so make sure you are aware of the legislation pertaining to your area; you can usually get this information from government and local government bodies.

Before purchasing stock, always ensure it is apparently healthy by giving it a thorough examination. Look for signs of mites or ticks on specimens. Choose only those with a sleek, unbroken skin, and ensure that the reptile is clear-eyed, alert, and plump. Ask if it is feeding regularly and on

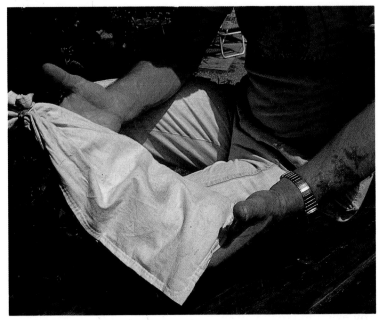

A simple bag like the one shown here is more than adequate for carrying most reptiles, but certainly not venomous ones. Photo by Susan C. Miller and Hugh Miller.

what. Examine the mouth and the vent for any sign of disease or abnormality.

TRANSPORT OF SPECIMENS

Most kinds of reptiles are transported in cloth bags, though aquatic turtles and crocodilians should be packed individually in insulated traveling boxes and preferably kept moist. Land turtles can be placed in a box loosely packed with clean hay or straw. Snakes and lizards are best placed individually in their cloth bags, a number of which may then be placed in a transport box. With heavier

MANAGEMENT OF CAPTIVE REPTILES

reptiles, ensure that each individual has its own compartment. Always avoid transporting in very cold weather if possible, otherwise arrange for some kind of temporary heating or insulation (plastic foam boxes come in handy here). Reptiles should reach their destination and be housed appropriately as soon as possible.

QUARANTINE

Always subject newly acquired reptiles to a couple of weeks of quarantine before introducing them to any existing stock you may have. By doing this you will help avoid introducing diseases into your stock. If the animal is still healthy after two or three weeks in an isolation cage, you can reasonably assume it is healthy and should then be safe to introduce to other stock.

HANDLING REPTILES

Different kinds of reptiles require different handling techniques. Many become very easy to handle once they are accustomed to it. Venomous snakes, however, should never be handled by amateurs at all; they should be left to the experts! Bites from small to medium non-venomous snakes may be painful but are not dangerous, though it is advisable to wipe any bites with a weak antiseptic solution to help avoid possible secondary infection.

Small non-venomous snakes up to about 50 cm (20 in) in length can be simply grasped gently but firmly around the neck with thumb and forefinger, with the body allowed to drape over the hand and arm. Larger (up to 150 cm, 5 ft) snakes require handling with both hands. If tame, the snake can be simply lifted by holding it about one-third and two-thirds the way along the body and draping it about the arms,

MANAGEMENT OF CAPTIVE REPTILES

restraining it as necessary. If aggressive or non-tame, the snake should be restrained just behind the head with one hand. Very large boas and pythons are dangerous if not tame and should always be handled with two people present. They are capable of giving deep lacerated bites and their powers of constriction should always be respected. Hand-reared boids usually become remarkably tame and can be handled with ease, even on reaching large sizes.

Small lizards like geckos, anoles, etc., may be simply grasped in the palm of the hand. They may try to bite, but these bites are usually too weak to cause any concern. Be gentle, and remember that many lizards will shed their tails if handled roughly. Larger lizards should be treated with a greater respect as some, particularly large monitors, iguanas, tegus, and similar types, have not only a powerful bite, but sharp claws and powerful whip-like tails as well. These are best restrained by the neck and tucked under the elbow to stop them from flailing about. Some specimens, especially those reared and handled frequently from juvenile age, soon become very tame and are then no problem at all to handle, even when fully grown.

Some turtles (especially snappers, some mud turtles, and softshells) are capable of giving serious bites, so these are always handled with extreme care. Throw a sack over vicious individuals before attempting to pick them up. Most other turtles and tortoises can be lifted simply by placing the hands under them and gripping the carapace. Crocodilians must, obviously, be treated with the greatest respect and are best left to the experts. Some hobbyists like to wear protective gloves when handling potentially dangerous species, but, in my opinion, these tend to make the whole aspect of handling more clumsy, especially if you intend to perform any minor operation or treatment on the animal.

NUTRITION

The nutritional requirements of captive reptiles have, until fairly recently, been poorly understood and mainly speculative, based on our knowledge of what the particular species is known to eat in the wild. There is some similarity between the the nutritional requirements of domestic fowl and reptiles, and some comparative studies have recently been made. A major variant, of course, is that reptiles do not have the high energy requirement of fowl to maintain a relatively high, constant body temperature.

The rise in popularity of reptiles as pets over the past few decades has led to some further research into the nutrition of these animals by a few scientists. Coupled with the great quantity of experiences documented over the years by amateur and professional herpetologists, this has led to a situation where no captive reptile should be deficient in any dietary requirement.

We know that to remain in the best of health and in prime condition all animals must receive a balanced diet. This consists of a number of primary nutritional constituents taken in appropriate amounts. These primary constituents are proteins (for the growth, repair, and replacement of body tissues as well as many other biological and metabolic functions); carbohydrates (for immediate energy requirements); fats (for stored energy requirements and insulation); minerals, particularly calcium and phosphorus (for bone growth and repair, proper function of the cell membranes, and the buffering of body fluids); vitamins (various functions) and finally, water —which is indeed the "elixir of life" for reptiles as well as us.

Unfortunately, different species not only acquire these dietary constituents in different ways, they may require them in varied percentages. It can be safely assumed that each reptile species in the wild obtains its balanced diet requirements by feeding upon the type and variety of food items available in its natural habitat. Remember that a species will have lived in that habitat for thousands (maybe millions) of years, and there will be a very close ecological relationship between it and the food available.

It will be almost impossible to provide our captive reptiles with the exact type and variety of foods they take in the wild, so we are thus obliged to provide substitutes that are nutritionally as similar as possible to the natural diet. Depending on the types of food they eat, reptiles can be categorized into three groups:

1) Carnivorous: Feeding on animal material (in its widest sense), ranging from small invertebrates to larger vertebrates depending on the size of the reptile. This is by far the largest group.

2) Herbivorous: Feeding almost exclusively on plant material (some lizards and tortoises only).

3) Omnivorous: Feeding on a fairly equal mixture of plant and animal materials.

These groups are far from clear-cut. A carnivore swallowing a whole prey animal will also be swallowing the contents of that animal's stomach, which may include a high proportion of plant material. Conversely, a herbivore will consume a fair proportion of the lower animal life that is associated with the plant material it eats.

MANAGEMENT OF CAPTIVE REPTILES 73

Any animal can bite, so why give it the chance? Illustrated here is the correct way to hold a non-venomous snake. Photo by Susan C. Miller and Hugh Miller.

74 MANAGEMENT OF CAPTIVE REPTILES

FOOD ITEMS

Perhaps the most important aspect of feeding captive reptiles is the consideration that "variety is the spice of life," though some species will, of course, require different general varieties of food from others. Before selecting a particular species of reptile to keep, first ascertain whether a

Looking for your own livefoods can sometimes be fun. This little boy is looking for, among other things, worms, beetles, etc., which can actually be fed to various reptiles. Photo by John Dommers.

MANAGEMENT OF CAPTIVE REPTILES

continuous supply of a variety of food items will be available. Also, if you do not relish the thought of feeding live foods to your reptiles, you should perhaps limit your choice to herbivorous tortoises or lizards. The general feeding requirements of various species are given with the species descriptions later in the text, but the following gives some general ideas on the types and variety of foods available.

COLLECTING LIVEFOODS

Although there are several kinds of livefoods that can be purchased at regular intervals or propagated in the home, the collection of a varied supply of invertebrates from the wild is highly recommended. This may be difficult if you live in the city and not very productive during the winter in colder climates, but it is worth making the effort; an hour or so spent during your summer weekend trip into the countryside can be very productive. A mixture of wild-caught invertebrates will not only provide your captives with a greater variety of diet, it will also help relieve boredom to a certain extent; insectivorous reptiles are likely to fast if continually given the same old items (mealworms and maggots, for example) to eat.

The most productive method of obtaining a selection of terrestrial insects and spiders is foliage sweeping. This is accomplished by sweeping the open mouth of a large, fine-meshed but sturdy net (a butterfly net probably is not strong enough) through the foliage of trees, shrubs, and tall grass. Such a sweep in the summer months will almost certainly catch quantities of beetles, bugs, caterpillars, grasshoppers, spiders, and so on, all items that most insectivorous reptiles will eagerly accept. The insects can be placed in a number of jars or small plastic containers for transport home.

Never release too many insects into the terrarium at any one time; let the reptiles consume what is available before adding more. Otherwise you will get escapes into the house or the excess insects will drown in the water and pollute the terrarium.

By searching under rotten logs, rocks, and other ground debris you have another good source of livefoods. Damp areas will produce large numbers of beetles, woodlice, and earthworms. By breaking open rotten timber, you may find the grubs of many insects. In some countries, termites are easily found; the soft, pale bodies of these "white-ants" make them an excellent and nutritious food for many small lizards and burrowing snakes.

In the warmer months, the flower garden is an excellent hunting ground for small insects. Tiny beetles and flies (especially useful for very small, juvenile lizards) congregate among the petals of flowers and can be collected using a "pooter." This consists of a glass or plastic bottle containing a stopper through which two glass tubes are passed. One of the tubes has a piece of rubber tubing about 15 cm (6 in) long attached to it. The insects are caught by placing the end of the rubber tube into the corolla of the flower and sucking sharply on the other tube with the mouth. The insects will be pulled through the long tube and fall into the bottle. A piece of fine mesh placed over the end of the mouthpiece tube will prevent unfortunate accidents.

Aphids (so-called greenflies and blackflies), the bane of the gardener in the summer, are another useful food item for very small lizards. These little insects are often found in great numbers on the new green shoots of domestic and wild plants. By simply breaking off a whole infested shoot and placing it in the terrarium, you will have a ready supply of food for your tiny lizards. Moths and other nocturnal insects can be caught in great quantities by using a light-trap. A white sheet is hung up in a suitable place and a strong light shone upon it. The insects will be attracted to the sheet and can usually be easily caught and placed in a container. A few specialized lizard species feed largely on ants, a food item that is rarely difficult to obtain, at least in the summer.

CULTURING LIVEFOODS

Mealworms:

These larvae of a species of flour beetle (*Tenebrio molitor*) were the standard staple diet of captive insectivorous creatures for a very long time until it was discovered that if fed alone they do not provide adequate nutrition. Although mealworms can be classed as "fairly nutritious," they should be used only as a part of a more varied diet. In particular, they tend to be deficient in calcium and certain vitamins, but a powdered vitamin/mineral supplement added to the insects will considerably improve their nutritive value. The fine grains of the powder will adhere to the bodies of the mealworms and will be taken in by the reptiles as they feed. In many countries, mealworms are readily available from specialist suppliers (pet shops selling birds often also sell mealworms), and they can be purchased in small quantities at regular intervals. Alternatively, having purchased your first batch, you can breed your own stock.

A number of mealworms is placed in a shallow tray or

Mealworms, like these *Tenebrio molitor*, are a good food source for some reptiles, but unfortunately are not nutritionally complete. Photo by Michael Gilroy.

box containing a layer (about 8 cm or 3 in deep) of bran covered with a piece of burlap. One or two pieces of carrot or apple (changed every couple of days) placed on the sacking will provide moisture for the insects. Best results will be obtained if you maintain your cultures at 26-31°C (79-88°F). Each month, a further culture is started with a few beetles from the first culture, until you have four cultures at various stages of development. The larvae in the first culture will pupate and emerge as adult beetles, ready to mate, lay eggs, and repeat the cycle. After about eight weeks, you will have a new generation of mealworms. Each month, a new culture is started from insects in the oldest culture, which is then discarded. By having four cultures at various stages of development, you will have mealworms of varying sizes, pupae, and adult beetles, all of which can be used as food for insectivorous lizards.

MANAGEMENT OF CAPTIVE REPTILES

Crickets:

In recent years cultures of these insects have become readily available through the pet trade both in the USA and Europe, and several specialist "cricket farms" have been started. They are a highly nutritious source of food for captive lizards, turtles, and small or juvenile snakes. Though there are many species of cricket, the one most commonly cultured is the common domestic cricket, *Gryllus bimaculatus*. The crickets may be kept in small containers (an old leaky aquarium is ideal) with rolls of corrugated cardboard or old egg boxes in which they can hide. A small saucer containing a piece of wet cotton wool will provide drinking water for the insects. They may be fed on bran and a regular small amount of fresh greenfood, carrot, etc. A small dish or two of damp sand or vermiculite should be provided for gravid females to lay their eggs. The dishes should be removed to a separate container at regular intervals and replaced with a new one. If kept at a temperature of about 26°C (79°F), the eggs will hatch in about three weeks. The newly hatched nymphs are about 3 mm (an eighth of an inch) long and are very suitable food for tiny hatchling lizards. There are four instars (nymphal stages), each one a little larger in body than the former, providing various sizes for your reptiles. The final adult size is about 15 mm (0.6 in). A carefully planned and harvested breeding colony of crickets will provide a constant supply of excellent livefood throughout the year.

To remove crickets from their container, pick up a piece of the cardboard in which they are hiding and shake the insects out into a jar. If placed in the refrigerator for ten minutes or so, the insects will be

80 MANAGEMENT OF CAPTIVE REPTILES

MANAGEMENT OF CAPTIVE REPTILES

subdued enough to prevent escapes when feeding them to your lizards.

Locusts:

These are also available from specialist suppliers and can be obtained in various instar sizes ranging from about 5 mm (0.2 in) to the adult size of 5 cm (2 in). Adult locusts are excellent food for a number of reptile species. They are a little more difficult to breed than crickets, though any problems can be quickly overcome. They can be fed on a mixture of bran and crushed oats, supplemented by fresh greenfood. Grass is a convenient greenfood that can be kept fresh by placing the stems in a bottle of water with wadding packed around the neck to prevent the insects from falling in and drowning. Locusts are best kept at a temperature of about 28°C (82°F) in a tall, well-ventilated aquarium tank or a glass-fronted box. The eggs are laid in slightly damp sand to a depth of 2.5 cm (1 in.), so suitable containers should be provided.

Flies:

Various species of flies are suitable food for lizards both in the larval and adult form. Fruitflies (*Drosophila* species for example), those tiny black insects that congregate and breed in ripe or rotting fruit, are an excellent food for the hatchlings of very small lizard species. A colony of fruitflies can soon be started (in the summer) by placing a box of banana skins or rotten fruit in a remote corner of the garden, where it will soon be teeming with flies. The little flies can be captured using a fine meshed net and transferred directly to the terrarium. Due to their fast breeding

---

Facing Page: Contrary to popular opinion, crickets are actually rather easy to breed in captivity. Photo by Michael Gilroy.

cycle, fruitflies are used extensively in genetics research. Laboratories culture them in jars containing an agar-based nutrient. Cultures are sold by biological supply houses and some pet shops.

The housefly (*Musca* species) is another excellent food for smaller lizards, while the bigger green- and blue-bottles (*Lucilia* and *Calliphora*) species may be suitable for larger lizards and some turtles. A selection of flies can be caught in the summer months by using a fly-trap. Such a trap can easily be made by constructing a 30 cm (12 in) cubical framework (from timber or wire) and covering this (except for the base) with a fine mesh (muslin or net-curtain material can be used). The framework is mounted on a flat wooden board with a 5 cm (2 in) hole in the center over which an inverted funnel is placed. The whole trap is placed in a suitable spot (preferably well away from the house) with the board standing on four bricks so that it is raised about 5 cm (2 in) from the ground. A piece of rotten fish or meat (the stronger smelling the better) is placed under the trap near the hole in the board. Many flies will soon be attracted to the bait and, if you occasionally give the board a gentle tap, the flies will panic and make for the nearest source of light—through the hole and funnel—to be caught in the trap. Flies can be extracted by having a muslin sleeve in one side of the trap, through which you can pass your hand and a jar. When not in use, the sleeve can be knotted at the end. Flies in jars can be subdued by placing them in the fridge for ten minutes or so, thus reducing the incidence of escapes when you add them to the terrarium.

Fly maggots are not generally regarded as good food for terrarium animals

Although somewhat difficult to culture on your own, fruitflies can be obtained through various means and are readily accepted by certain turtle and lizard species. Photo by Michael Gilroy.

as they have a very tough skin and collected larvae may contain toxic substances. Gentles, the larvae of blue-bottle flies, are occasionally available in bait shops. These are purged by placing them in bran or clean sawdust for a few days. The gentles themselves may be fed in small quantities to medium-sized lizards or turtles, but it is far better to allow the maggots to pupate and develop into adult flies, which are more nutritious and digestible.

Earthworms:

These are an excellent nutritious food for those reptiles that will accept them. Cultures of various earthworm species are available from bait shops and some pet shops, though if you have a garden it is quite easy to find your own supply. A good method of obtaining a fairly continuous supply of earthworms during the warmer months is to clear a patch of soil in some corner of the garden, preferably where it is shaded from the sun. Place a 5 cm (2 in) thick layer of dead leaves or grass clippings over the patch and cover this with a piece of burlap sacking (pegged down to prevent it from blowing away). This should be kept moist (but not wet) by regular hosing. Earthworms will soon congregate in the rotting vegetation below the sacking and can easily be collected.

Other Invertebrate Foods:

The quest for variety in feeding captive terrarium animals has led to many recent experiments into the culture of suitable food species. Many species may be purchased and/or bred, and it is up to the individual how much effort he is prepared to devote to feeding his lizards. Other invertebrates that may be available as cultures include whiteworms, snails (various species), wax moths, flour moths, grain beetles, weevils, cockroaches (various species), and stick insects. Purchased cultures are usually offered with instructions on how to further proliferate them. Aquatic invertebrate foods that are relished, particularly by freshwater turtles and some water snakes, include shrimp, crayfish, and similar items.

Mice and Rats:

Most of the larger lizards, snakes, turtles, and crocodilians will feed readily on mice or rats. The availability of laboratory mice and rats makes these animals an almost staple diet for many captive reptiles. Reared on a balanced diet, rodents themselves are a balanced diet for our larger carnivorous reptiles. A self-sufficient colony of rats

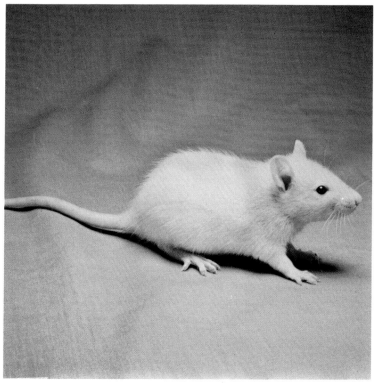

Rats seem to stir up many adverse reactions, but in reality the domestic variations are quite easy to breed and make an excellent source of nutrition. Photo by Dr. Herbert R. Axelrod.

and/or mice is quite easy to maintain but somewhat time consuming. If you do not want to take the trouble to do this, mice and rats may be purchased as necessary from pet shops that cater to reptiles; they may even supply them in a deep-frozen form that can be

MANAGEMENT OF CAPTIVE REPTILES

This is a nest of newborn rats, otherwise known as "pinkies." Photo by Susan Fox.

fed to the reptiles after being thoroughly thawed out.

Chickens:

The young chicks of domestic fowl may be obtained inexpensively from hatcheries either alive or dead (sometimes deep-frozen, or you can deep freeze them yourself). Chicks that have started to grow and are a few days old are more nutritious than hatchlings. Day-old chicks can be bought inexpensively and reared to sizes appropriate as food for the reptiles you keep. Local laws may prevent the maintenance of chickens and other fowl in residential areas; check first.

Other Vertebrate Foods:

Some reptiles (perhaps unfortunately) feed on amphibians and other reptiles—certain small snakes feed exclusively on lizards and/or frogs and it is often difficult to get them onto any substitute. By rubbing a pinkie mouse with the skin of a frog, you can sometimes deceive the snake with scent. If lizards and frogs must be captured purely as food for other reptiles, be sure that only common and non-vulnerable species are used. Live fish can be used for some reptiles. I have often gotten lizard- or frog-eaters to take guppies or small goldfish by placing the fish in a container of water so shallow that they flip about and attract the reptile's attention. Of course, many water snakes and turtles will take fish in their own right.

Vegetable Foods:

Some lizard species, especially iguanids, and many land tortoises are at least partially herbivorous and will take varying amounts of vegetable matter. Even some species generally considered to be carnivorous (certain skinks, lacertids, freshwater turtles, etc.) will occasionally take soft fruit or grated vegetables. Most of the usual domestic fruits, salad greens, and vegetables are suitable. By experimenting with a variety of foods a feeding strategy can be formulated. Lettuce, cabbage (shredded), carrot (grated), tomatoes, peppers, cucumbers, apples, pears, bananas, grapes, oranges, boiled potatoes, and various other items can be tried. You can even try some of the many nutritious weeds of the kind collected by rabbit fanciers (dandelions, chickweed, shepherd's purse, and so on). Some particularly fussy feeders may require a whole range of experimentation before you reach a satisfactory feeding regimen. Fickle feeders are often tempted with such items as fresh strawberries or even canned peaches or canned peas. Species such as the Desert Iguana, *Dipsosaurus dorsalis*, which feed on pungent desert plants in the wild may often be tempted with plants such as mint or rosemary from the herb garden.

Other Foods:

Crocodilians, some carnivorous lizards, and

most turtles, especially those species which feed regularly on carrion in the wild, will eagerly accept pieces of lean meat, heart, or liver (minced for the smaller individuals). If mixed with raw egg (including the crushed shell), the meat may be even more eagerly accepted. Canned dog or cat food may also be accepted by many species. Some large monitor lizards will swallow whole chicken eggs, shell and all, while smaller ones can be given pigeon or quail eggs. Snakes generally will not take lean meat but must be given whole prey animals.

Supplements:

Lean meat should never be fed as a staple diet as it lacks the important roughage of bone, fur, or feather. If it becomes necessary to give lean meat for extended periods (due to a shortage of mice, rats, or chicks for example), then a suitable vitamin/mineral supplement must be added. Powdered supplements can be quite easily added to the meat and scoured into it. Ask your veterinarian or enquire at your pet shop as to which brands are most suitable. Be sure that a range of vitamins is available, as well as minerals, especially calcium and phosphorous compounds.

It is advisable to give supplements to all insectivorous and herbivorous reptiles on a regular basis (say twice per week) by sprinkling the powder onto the food.

FEEDING STRATEGIES

Small reptile species and juveniles of larger ones generally feed more often than large ones. Small insectivores and all herbivores should be fed daily, while most of the larger carnivorous lizards will get by on two or three sizable meals per week. While young snakes will

MANAGEMENT OF CAPTIVE REPTILES

Some people find the fact that snakes eat mice repulsive, but this is simply a part of their natural behavior. Shown is a juvenile Black Rat Snake, *Elaphe obsoleta obsoleta*, taking a pinkie mouse. Photo by William B. Allen, Jr.

require two meals a week, this will drop to one meal per week and eventually one per fortnight as the snake matures. It is difficult to make any hard and fast rules regarding quantities. You should aim at keeping the diet balanced without overfeeding (obesity is a common cause of premature death in captive reptiles). A certain amount of experimentation will be required before you arrive at a suitable routine.

Health and Hygiene

The words health and hygiene complement each other. Good health can be defined as a state of physical and mental well-being, devoid of stress and disease. Hygiene itself is the science of disease prevention. If we keep animals confined, it is our responsibility to apply hygienic practices to the extent that those animals are always in good health. Having acquired healthy specimens that have undergone a period of quarantine, it is our duty to keep them that way.

CLEANING

To prevent or at least to minimize the risk of an outbreak of disease, terraria must be maintained in a spotlessly clean condition, but try to evolve a routine in which the animals are not stressed by over-disturbance. For the simple "clinical" type of terrarium, absorbent paper may be used as a substrate as this can easily be changed each time it becomes soiled. With other forms of substrate, such as sand or gravel, fecal pellets can be removed using a scoop or small shovel. About once per month the whole terrarium should be cleaned; materials should be removed and either discarded or scrubbed clean. The interior of the terrarium and its contents should be scrubbed with warm soapy water and a mild disinfectant such as household bleach or povidone-iodine, then thoroughly swilled out with

HEALTH AND HYGIENE 91

There is nothing wrong with giving your reptile a quick bath outside, much like you would a dog, ferret, or a few other pets. Shown here is a young (and very patient) Yellow-footed Tortoise, *Geochelone denticulata*. Photo by Susan C. Miller.

clean water before being dried and refurnished.

During cleaning operations the reptiles can be placed in a spare cage or a plastic box. A plastic trash can is useful for large species. Water for drinking and/or bathing should be changed very regularly, preferably daily or more often if necessary, unless of course you have a self-cleansing filtration system installed. The glass viewing panels should also be kept crystal-clear, for esthetic purposes as well as those of hygiene.

HIBERNATION

The importance of hibernation for many reptile species cannot be over stressed. Reptiles from temperate and subtropical climates usually hibernate for varying periods of time, depending on the ambient temperatures. In the wild, reptiles seek out hibernation spots far enough below ground to be unaffected by frost, and many species can withstand quite low temperatures providing they do not freeze. The period of hibernation is a part of the life cycle that helps bring some reptiles into breeding condition. In the past many captive reptiles were kept active in high temperatures throughout the year. The general opinion today is that non-tropical reptiles should be given varying periods of artificial hibernation to enhance their lives and increase the prospects of breeding. A hibernating "pet" reptile may seem boring, as it will not be seen for quite a long period. However, a short "rest period" at lower temperatures seems to be an adequate substitute for full hibernation.

Only healthy, well-fed specimens should be hibernated. Stop feeding and gradually reduce the temperature over a period of several days. The minimum

HEALTH AND HYGIENE

temperature will vary from 4°C (39°F) for cold-temperate species to 10°C (50°F) for sub-tropical species. The photoperiod can also be reduced at the same time. The animals should be kept at these temperatures in an unheated but frost-free room for periods ranging from 4-12 weeks depending on their natural climatic zones. After the "rest-period," the reverse procedure should be used to bring the temperature and photoperiod back to "summertime."

DISEASES AND TREATMENT

Kept hygienically in a stress-free environment and receiving an adequate diet, most reptiles will remain resistant to disease. Many outbreaks of disease can be related to some inadequacy in the captive management, so careful thought must be applied at all times in providing their needs. If it is suspected that all is not well with your animals, a veterinarian should be consulted. Though many veterinarians are inexperienced in reptile medicine, most will be able to put you in touch with one who is. Some of the more usual problems are detailed below.

Nutritional Problems:

A common cause of sickness in many captive reptiles can be put down to a deficiency in certain dietary constituents. With a variety of the right types of food, vitamin and mineral supplements, fresh water, and an opportunity to bask in sunlight or artificial sunlight, the incidence of such conditions will be minimized.

Wounds and Injuries:

Though not strictly diseases, wounds caused by fighting, attempting to escape, etc., are susceptible to infection and must be treated. Shallow wounds will usually heal automatically if swabbed daily with a mild antiseptic such as povidone-iodine. Deeper or badly infected wounds should be treated by a veterinarian as in some cases surgery and suturing may be required.

Ectoparasites:

These are parasites that attack external areas of the body in order to suck blood. Ticks and mites are the most usual external parasites associated with reptiles. Ticks are often found attached to newly captured specimens and may range up to 5 mm (0.2 in) in length. They fasten themselves with their piercing mouthparts to the reptile's skin, usually in a secluded spot between scales, often around the vent, below the neck, or where the limbs join the body. Do not attempt to pull a tick directly out, as its head may be left embedded in the skin, causing infection later. The tick's body should first be dabbed with a little alcohol or petroleum jelly to relax the mouthparts. The tick can then be gently pulled out with forceps. Once all ticks are removed from a wild-caught specimen, a further infestation in the terrarium is unlikely. Ticks found on reptiles are not likely to carry Lyme disease.

Mites are more serious as they can often multiply to large numbers in the terrarium before they are even noticed. They do not necessarily stay on the reptile's body all of the time, but may hide in crevices in the terrarium. In great numbers, mites can cause stress, anemia, molting problems, loss of appetite, and eventual death; they are also capable of transmitting blood-borne pathogenic organisms from one reptile to another. The individual reptile mite is smaller than a pinhead, roughly globular in shape, and grayish in color,

HEALTH AND HYGIENE

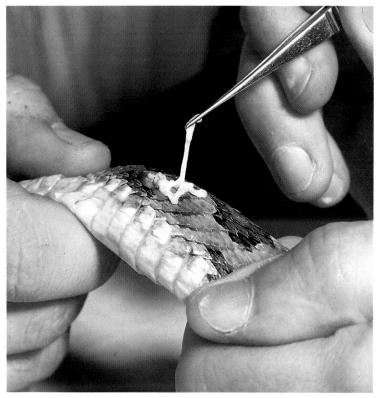

Although some medical procedures may seem simple, almost all should be handled strictly by a professional veterinarian. Photo by S. Kochetov.

becoming red after it has partaken of a blood meal. In a heavily infested terrarium, the mites may be seen running over the surfaces, particularly at "lights-on" in the morning, and their silvery, powdery droppings may be seen on a reptile's skin. Mites are most often

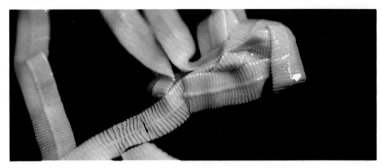

The tapeworm shown here is a good example of a common endoparasite. Photo by William B. Allen, Jr.

introduced to the terrarium with new stock (another good reason for quarantine and careful inspection).

Fortunately, mites can be quickly eradicated using an insecticidal strip of the type used to control houseflies. A small piece (the size of a thumbnail) of such a strip placed in a perforated container and suspended in the terrarium will kill off free-moving mites. Remove the strip after three days then repeat the operation ten days later to kill off any newly hatched mites. Two or three treatments will usually destroy all mites in the terrarium. Do not leave the strips in the cage for extended periods in case they become harmful to the reptiles themselves. There is considerable doubt as to the safety of the vapona insecticide used in such strips. Other methods of mite control include injections that your veterinarian may be able to help you with in extreme cases.

Endoparasites:

These are organisms that live inside the body, usually in the alimentary tract. The ones with which the reptile

HEALTH AND HYGIENE

The moment you spot a tick, you should remove it immediately. Be careful not to let the head break off and remain in the body. Photo Isabelle Francais.

keeper is most likely to be concerned are roundworms and tapeworms. Wild reptiles are nearly all infected with worms of one form or another, but in most cases there is no danger to the reptiles. However, during times of stress (capture for example), normal resistance to the worms may be lost, triggering a massive increase in size or numbers of worms and leading to anemia, general lethargy, loss of appetite, and eventual death. Routine microscopic examination of fecal samples in a veterinary laboratory will reveal infestations. There are several proprietary brands of vermicides available through your veterinarian that may be offered with the food or, in severe cases, via stomach tube.

Bacterial Infections:

There are many forms of bacterial infections that can infect reptiles, especially in unhygienic conditions. Infective salmonellosis is an intestinal disease that is known to be transmitted from reptiles to man (especially from freshwater turtles), so it is important to thoroughly wash the hands after each cleaning or handling session. In lizards and snakes, infective salmonellosis may be indicated by watery, green-colored, foul-odored feces. The most commonly encountered bacterial diseases in reptiles are usually caused by gram-negative bacteria such as *Pseudomonas* or *Aeromonas*. Consult a veterinarian, who will probably treat the infection with antibiotics.

Protozoan Infections:

Various enteric infections

Ideally, a shedding should be smooth, and the old skin unbroken. The shed from this stunning Gray-banded Kingsnake, *Lampropeltis alterna*, is basically that way. Photo by K. T. Nemuras.

HEALTH AND HYGIENE 99

can be caused by protozoa, including *Entamoeba invadens*. If untreated, the disease can rapidly reach epidemic proportions in captive reptiles. Symptoms include watery, slimy feces and general debilitation. Treatment with metronidazole via stomach tube (under veterinary direction) has proved effective for this and other protozoan infections.

Skin Problems:

A common cause of skin infections in lizards and snakes is the inability to molt properly, often as a result of a mite infestation or stress brought about by various other factors. Mite infestations should be cleared immediately and aid should be given to reptiles experiencing difficulty in shedding. Most healthy reptiles will molt their skins without any problems several times per year, a natural phenomenon related to growth. The skin is normally shed in patches in lizards and the whole process should take no more than a few days. In snakes the molt is heralded by a milkiness in the eyes, and the skin should be shed in a single piece in a relatively short time. The snake uses rough objects to loosen the old skin around the lips, then literally crawls out of the old skin, which turns inside-out. In sick snakes the molt may be patchy and parts may refuse to budge. Disease organisms can grow behind persistent patches of old skin that do not come off readily. Problem skin can usually be loosened by placing the reptile in a bath of very shallow, warm water for an hour or so. If it does not float away in the water, it can be very gently peeled off with the fingers.

Other infections of the body surface may include abscesses, which appear as lumps below the skin. These are usually caused by infection building up in the

HEALTH AND HYGIENE

After receiving a nasty wound from a fight with a cagemate, this Reticulated Python, *Python reticulatus*, was expertly sutured and eventually healed completely. Photo by William B. Allen, Jr.

flesh after the skin has been accidentally damaged for one reason or another. Abscesses should be referred to a veterinarian, who will give antibiotic treatment. In severe cases the abscess may be surgically opened, cleaned up, and then sutured.

Respiratory Infections:

Though relatively uncommon in most reptiles, respiratory infections may occur occasionally in stressed specimens and especially post-hibernating tortoises. The patient will have difficulty in breathing, the nostrils will be blocked,

This ugly head wound penetrating the skull was acquired through a fight with a mate during breeding time. Photo of the Common Iguana, *Iguana iguana*, by William B. Allen, Jr.

and there will be a nasal discharge. Often the symptoms can be alleviated by washing away the discharge with a mild antiseptic solution and moving the patient to a warmer, drier, well-ventilated terrarium. More serious cases will require antibiotic treatment from the veterinarian.

Captive Breeding

REPRODUCTIVE CYCLES

Unlike many mammals and birds with their constant, relatively high normal body temperatures, reptiles are unable to adapt to climates significantly different from those of their natural habitat. No captive breeding will occur unless you attempt to reproduce these climates as closely as possible. The reproductive cycles of all species are to a greater or lesser extent affected by environment and seasonal changes. Before attempting captive breeding, it is necessary to try and provide a terrarium microclimate compatible with the species; the provision of the correct "seasonal changes" in the terrarium will greatly enhance the chances of successful breeding.

On reaching maturity, most reptiles from temperate climates breed only once a year, while some of the smaller species from sub-tropical and tropical climates may breed twice or more. Temperate species have a relatively short time to complete their breeding cycle, so courtship activity must start as early as possible in the season. The period of hibernation plays an important part in preparing such species for breeding, thus spring is commonly regarded as the mating season. In tropical species the breeding cycle may be affected by the onset of "wet" or "dry" periods; some may benefit from a period of estivation or changes in humidity. By using a combination of heaters, thermostats, lamps, timers, and humidifiers, it is possible to reproduce a mini-climate suitable for the reproductive cycles of most species.

With the possible exception of some turtles and "social" lizards (not taking into the account the occasional "aggregations" of hibernating reptiles that are not necessarily being social, but using the few available and ideal hibernation sites), most reptiles live fairly solitary lives outside the breeding season, though there may be fairly active territorial interaction. Breeding responses are thus more likely among non-social reptiles if the sexes have been kept separately until the right time for a breeding response arises. Some species live in "gregarious colonies" (*Agama agama*, for example), often in a relatively small area such as a single tree or rocky outcrop. The strongest, or at least the most aggressive, male will adopt the highest point available and advertise his superiority by taking on bright colors and making bizarre body movements including head-bobbing, dewlap-extending, or other activities depending on specific behavior. The dominant male will usually have the choice of females, but he will have a hard job keeping competitors at bay, often temporarily losing his superior position to one interloper while defending it from yet another.

A sexually aroused male lizard will often contort his body into bizarre shapes, ostensibly to make himself look more attractive to a female. With arched or twisted body, outstretched limbs, and quivering tail, he will approach the female who, more often than not, is quite unimpressed by his performance and may even take flight. The male will grab the female in the region of the neck, and if she is receptive she will submit; if not, she will fight him off. Having secured a receptive

CAPTIVE BREEDING 105

Shown here is a wild pair of breeding Green Anoles, *Anolis carolinensis*. Photo by R. Allan Winstel.

female, the male will maneuver his vent into apposition with hers before inserting one of his hemipenes into her cloaca. Copulation can take minutes or hours, depending on the species and the circumstances. Lizard mating can appear to be excessively violent, but individuals are rarely injured seriously, and they should not be disturbed during the process.

Snakes are relatively more gentle when engaged in courtship and mating, but rival males will sometimes indulge in minor wrestling matches in which each will rear up and try to force the opponent to the ground. Once this has been done, the vanquished will depart, leaving the victor to

A male snake will often grasp the female's head lightly to ensure that she does not move off during copulation. Photo of two Central Plains Milk Snakes, *Lampropeltis triangulum gentilis*, by L. Trutnau.

CAPTIVE BREEDING

Although not for amateurs, tortoise breeding can be very rewarding. Photo of *Geochelone carbonaria* by Harald Schultz.

mate with the available female. The male crawls slowly along beside the female, nuzzling her body with his snout. If she is receptive, she will stay relatively still and the male will entwine his body around hers until their vents are in apposition and he is able to insert one of his hemipenes into her cloaca.

Crocodilians are rarely bred in captivity, other than in zoos or special refuges.

Courtship is often accompanied by much noise and apparent violence, but serious injuries among the rival males are rare. The female lays her eggs in a mound built from a mixture of soil and vegetation that she scrapes together. She stays on guard at or near the nest site until the young hatch and then transports them safely and gently to the water. Crocodilians may be very dangerous and require immense amounts of space in which to breed, and are thus really beyond the scope of this book.

Turtles have various means of courtship. Depending on the species, it can range from apparently violent butting (the male butting the female's shell with his) and biting in land tortoises, to delicate digital signaling in the case of many freshwater turtles. Turtles copulate in the more conventional way of the male mounting on the back of the female.

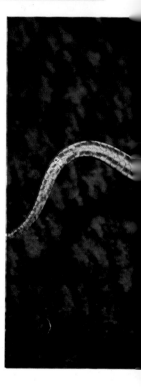

No, this is not a pencil! It is called a sexing probe, and can be used to determine the gender of almost all snakes. Photo of the proper method of usage by Jeff Gee.

SEX DETERMINATION

If you want to breed reptiles, one of the most obvious requirements should be that you have at least one male and one female of the same species (there are a very few species that have been proven to be parthenogenetic, females

CAPTIVE BREEDING 109

being capable of producing offspring without being fertilized by a male). Most snakes are fairly difficult to sex, being very similar in appearance. In many species the tail of the male is longer and more tapering than that of the female. The male's tail usually has a bulge at its base housing the inverted hemipenes. Many modern herpetologists determine the sexes of snakes by using sexing probes. These are made from stainless steel or a hard synthetic material and

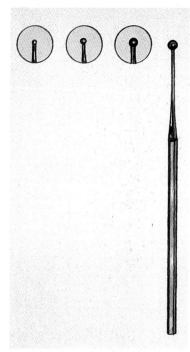

Sexing probes will of course vary in size and obviously should be used in relation to the size of the serpent in question. Artwork by Larry Nelson.

lubricated (use petroleum jelly or mineral oil) ball-tip into either side of the vent and pushing gently in the direction of the tail. In males, the probe will go a considerable distance inside the inverted hemipenis, while it will only pass a short distance into the cloaca of the female. This is a fairly skilled task and should only be carried out by an expert.

It is relatively easy to distinguish the sexes of most lizard species because of differences in color, size, behavior, tail length and shape, and so on. However, there are some species that show no obvious sexual dimorphism. When dormant, the paired hemipenes of the male are inverted into cavities in the tail base. In many species this gives males an obviously swollen region in the base of the tail that contrasts strongly with the thinner tail base of the female. Sexing probes have

basically resemble knitting needles but with a small ball at the tip rather than a point. Obtainable in various sizes from specialist suppliers, the probes are used by inserting the

Note how round the eggs of the Radiated Tortoise, *Geochelone radiata*, are. Photo by R. D. Bartlett.

been used with some success in certain lizard species.

Turtles also vary in their degree of easiness to determine sexes. In certain cooters and sliders the males are smaller, more colorful, and may have extremely long claws on their forelimbs. With many turtles, the male's tail (which contains the withdrawn penis at its base) is much thicker than that of the female. With land tortoises and some turtles, the plastron of the male is often concave (probably to make it easier for him to mount the female during copulation); that of the female is flat or even slightly convex.

EGGLAYING AND BIRTH

Though most reptiles lay eggs (oviparous), some species produce fully formed young at birth. In the majority of cases the latter is a result of ovoviviparity (eggs developing full-term in the maternal body) in which the eggs hatch just prior to or during deposition. Gravid females actively seek out suitable spots in which to lay their eggs and seem to have an instinctive insight into the exact conditions required for hatching. Warmth and humidity seem to be the most important factors in incubation, and eggs are usually laid in excavations in a sunny spot but with a moist substrate. An exception to this is geckos, which lay their fragile, adhesive eggs behind tree bark or similar structures. Concealment of the eggs from predators seems to be a prime consideration, as anyone who has tried to find wild reptile "nests" will tell you.

In an advanced state, gravid female lizards and snakes take on a very plump appearance and the eggs may be seen and palpated as a series of bulges on either side of the abdomen. The period from mating to

CAPTIVE BREEDING 113

The moment of truth: a hatchling Eastern Milk Snake, *Lampropeltis triangulum triangulum*, emerging from the egg. Photo by William B. Allen, Jr

egglaying varies from species to species and also depends on various environmental factors. It may range from 30 days in oviparous species to 100 days or more in ovoviviparous species.

Gravid females should be given facilities for laying eggs in the terrarium. This can consist of trays filled with damp sand for lizards or plastic boxes stuffed with moist sphagnum moss for snakes. Many reptiles may

CAPTIVE BREEDING

decide that what you have to offer them is inadequate and may end up dropping the eggs indiscriminately over the substrate. Wherever the eggs are laid in the terrarium, they are more likely to hatch if removed and artificially incubated. Once eggs are laid, the majority of species show no further interest in their offspring. Some pythons are notable exceptions. The females of a few species of python lay eggs in a mound and coil about them, protecting and warming them for the period of incubation (the incubating python is one of the few reptiles that can raise its body temperature by physiological means). Once the eggs hatch, however, she will lose interest in her "brood." Unlike most reptiles, captive pythons are usually more successful in hatching if allowed to brood their own eggs.

INCUBATION

The eggs of most reptile species (with the exception of some turtles,

crocodilians, and geckos that have brittle-shelled eggs) have a soft, white, leathery shell designed to absorb moisture from the substrate or incubation medium. Newly laid eggs often have dimples or collapsed areas, but these will soon fill out as moisture is absorbed. The eggs should be carefully removed (they should be kept the same way up as they were found and preferably not be "turned" regularly as you would the eggs of birds) and partially buried in an incubation medium contained in a shallow container. For convenience, the eggs can be laid in neat rows and buried to about two-thirds of their thickness. The top third of each egg is left exposed so that it may be inspected with the minimum of disturbance.

Many kinds of incubation media including sand, earth, peat, absorbent paper, and so on have been successful, but, in my experience, the

Below is a closeup of two Sinaloan Milk Snakes, *Lampropeltis triangulum sinaloae*, with their tails entwined during copulation. Photo by L. Trutnau

CAPTIVE BREEDING

Simple glass bowls with a layer or two of moist paper-toweling are perfectly acceptable for incubating large numbers of reptile eggs. Photo by William B. Allen, Jr.

most satisfactory method has been the use of granular vermiculite. This inert, sterile, absorbent insulating material is available in various grades. For general incubation purposes a fine grade is used. Mixed with about its own weight of water, the vermiculite will provide an ideal medium for incubation. The lid of the incubation box should have a few ventilation holes to allow for air circulation. The box containing the eggs is placed in an incubator and maintained at a suitable temperature. Adequate hatching temperatures for most species are in the range of 25-30°C (77-86°F). Fluctuations within this range usually are quite acceptable and seem essential in some cases for the development of a more or less equal (50/50) sex ratio.

Any kind of incubator that can maintain the hatching temperature range will do. A simple wooden box containing an incandescent light bulb and a thermostat to regulate the temperature is often quite adequate, though more sophisticated appliances are available from specialist suppliers.

With a thermometer in the box you will be able to monitor the temperature. It is best to use a blue bulb or to mount the bulb in some sort of cover to minimize light intensity. Heating pads, cables, porcelain heaters, or an aquarium heater in a jar of water may also be used to good effect.

As the eggs develop, they absorb moisture from the surrounding medium and

Another, slightly more popular and reliable, method of incubating eggs is the usage of moistened vermiculite in a plastic shoebox. Photo by William B. Allen, Jr.

CAPTIVE BREEDING

Hatchlings like these popular corn snakes, *Elaphe guttata guttata* (shown here in both normal and albino form), will very probably take their first meal without fuss. Photo by K. T. Nemuras

increase in weight. Infertile eggs will not absorb water, but do not discard them until you are certain they are dead. Incubation times will vary from species to species and may be anything from 30 to 100 days or more. This incubation time can be frustrating, and you may have a frequent desire to inspect the eggs, even open one up to see if an embryo is developing. Patience is certainly a virtue here, however, and eventually you will hopefully be rewarded with a batch or two of lively hatchlings.

A sharp projection on the snout, known as an egg

CAPTIVE BREEDING 119

Somewhat rounder than snake eggs are those of turtles. Shown here are fourteen from a softshell turtle of the genus *Trionyx*. Photo by Bob Gossington.

tooth, is developed by most late embryo reptiles. This is used to slit open the tough, parchment-like shell at the time of hatching, but is shed shortly afterward. Sometimes hatchlings seem to take a long time to emerge, often 24 hours or more, but the temptation to "help" them is usually best suppressed unless the reptile is having obvious difficulties. Occasionally the egg fluids will dry out too quickly and harden, causing

CAPTIVE BREEDING

the little hatchling to adhere to the egg shell. This can be overcome by gently dabbing the affected parts with a piece of wadding soaked in lukewarm water.

REARING

The rearing of some of the smaller species constitutes a challenge, the main problem being the maintenance of a constant supply of small insects for small lizards and snakes. As soon as hatchlings are free-moving and detached from the egg shell, they should be removed from the incubation chamber and placed in "nursery" accommodations. This should be simply furnished, but should have all the necessary life-support systems. Small species can be housed in small plastic tubs or boxes with ventilation holes drilled in the lid; a number of tubs can then be placed in a larger, heated terrarium. Do not attempt to remove the yolk sac, which will soon shrivel up and drop off, leaving a tiny scar on the belly.

Given suitable conditions and an acceptable food supply, most youngsters will start to feed shortly after hatching or, in the case of snakes, after the first shed, which usually occurs within 2-3 days of hatching. If one kind of food is ignored, keep trying others until you are successful. Once a youngster starts taking one kind of food it will not be long before it is prepared to try others. It is a good idea to weigh your specimens regularly and monitor their growth progress. In any case, it is wise to keep records of the complete progress of your reptiles, both for your own use and for that of others in the future.

Facing Page: Easy to raise but unfortunately very rare are these tiny Bog Turtles, *Clemmys muhlenbergii.* Photo by R. T. Zappalorti.

CAPTIVE BREEDING 121

Section Two
A SELECTION OF SPECIES

This extended section deals with a selection of species of the types of reptiles most likely to be kept by the beginner or the amateur herpetologist. In view of the large numbers of species, it is possible to describe only a token number here. Readers interested in studying further examples are urged to refer to some of the excellent field guides now available. Species that are unlikely to be kept in the home, including crocodilians, marine turtles, the tuatara (this rare, protected creature will probably never be available on the pet market), and venomous snakes (which should never be kept by beginners or amateurs) are not included in detail.

Tortoises and Turtles
Order Chelonia

All tortoises and freshwater and marine turtles belong to the reptilian order Chelonia, which includes two suborders, about 10 families, and some 250 species ranging is size from small 15 cm (6 in) freshwater turtles to giant land tortoises of the Galapagos in the Pacific and Aldabra in the Indian Oceans and huge marine turtles including the massive Leatherback Turtle or Luth, *Dermochelys coriacea*, adults of which may have a shell length of 180 cm (6 ft) and a weight of 500 kg (over half a ton)! Chelonians are probably the best loved of the reptiles and are often the first "pets" kept by budding herpetologists. They are characterized by the shell, composed of a carapace above and a plastron below, which encloses the body in armor. Types of turtles kept by hobbyists include some of the smaller land tortoises and freshwater turtles.

SUBORDER CRYPTODIRA

Spur-Thighed or Greek Tortoise
Testudo graeca

True tortoises belong to the family Testudinidae. Although it has the specific name "graeca" and is thus sometimes called the Greek Tortoise, this species is native not to Greece but to most of northern Africa and

Facing Page: Rarer than rare, pictured here is an albino Desert Tortoise, *Gopherus agassizi*. Photo by Ken Lucas, Steinhart Aquarium.

to the southwestern part of the Iberian Peninsula. It also has subspecies found from Israel through Turkey and southern Russian states to central Iran. Adults reach about 30 cm (12 in) in total length. It can be readily distinguished from closely related species by the presence of a bony spur on the rear of each thigh. The shell color is a mixture of black, browns, and buffs, and the skin of the head, neck, limbs, and tail is a fairly plain grayish brown. At one time, this and other European species were imported into northern Europe in great numbers for the pet trade, so much, in fact, that wild specimens were becoming endangered. Most wild populations are now protected and hobbyists are supplied with captive-bred specimens, which can be very expensive.

This species is mainly herbivorous and should be fed on a wide variety of greens, fruits, and vegetables. It will also take a little meat in the form of raw minced beef or dog or cat food. A good diet is essential to the health of captive tortoises. Although they may be kept outdoors and will even attempt to breed in a sun-oriented enclosure in cooler climes, their eggs will not hatch unless taken for artificial incubation. It should be allowed a period of hibernation in winter. A closely related species with similar habits and care is Hermann's Tortoise (*Testudo hermanni*), which ranges from the Balkan Peninsula through Italy to Iberia. Though very similar to *T. graeca*, it lacks the spurs on the thighs and instead has a hard spine at the tail tip. The rare Margined Tortoise (*T. marginata*) from the mountain slopes of Greece is characterized by the edge of

Facing Page: Now a common hobbyist's choice, the Starred Tortoise, *Geochelone elegans*, is often bred in captivity. Photo by K. T. Nemuras.

the shell being strongly flared in adults.

Gopher Tortoise
Gopherus polyphemus

A typical North American tortoise that was once fairly abundant in the southeastern part of the USA, the Gopher Tortoise is so named because of its habit of sharing its deep burrows with gophers and other burrowing animals. Reaching an average adult length of 25 cm (10 in), the basic color of the shell is blackish brown with a blotchy pattern of lighter browns. Its head and limbs are grayish brown. Due to over-collection for food as well as for the pet trade, this species is much scarcer than it should be and it is fully protected in most states. In the wild they are nocturnal, spending the daylight hours in their burrows, but they may become semi-diurnal in captivity. They require a diet of mixed greenfood, fruit, vegetables, and a small amount of animal matter. An occasional vitamin/mineral supplement will not go amiss. The Desert Tortoise (*G. agassizi*) is a closely related species from the American Southwest that requires a dryer environment and, in view of its specialized requirements, is not really suitable as a pet.

Leopard Tortoise
Geochelone pardalis

This is a tropical species native to much of Africa south of the Sahara. In spite of its large range, it is nowhere abundant and is thus only occasionally available, though it is being bred in captivity. It has an attractive spotted shell (which gives it its common name), and adults reach a maximum length of 50 cm (20 in). In captivity, these tortoises require heated cages with perhaps a slight reduction in winter temperature, but hibernation is unnecessary. They will feed on a diet similar to that described for the Spur-thighed Tortoise, with perhaps an emphasis on the fruit.

Starred Tortoise
Geochelone elegans

This is one of the most attractive tortoises because of the yellowish star-shaped patterns on a dark background on each of its dorsal scutes. Reaching a maximum length of 25 cm (10 in), it occurs naturally in India and Sri Lanka. Habits, care, and feeding are similar to that of the Leopard Tortoise.

Red-legged Tortoise
Geochelone (Chelonoidis) carbonaria

This is probably the best known of the South American tortoises, and it is found in a range of habitats from Panama to Paraguay. It has a dark, almost black shell, with the center of each carapace shield and the edges of the marginal shields being a bright reddish brown. Scattered scales on the limbs and head are also brownish to yellowish red. Reaching a maximum of 40 cm (16 in), this species is occasionally available. It requires a heated terrarium and medium humidity. Being omnivorous, it requires a varied diet of fruit and meat (dog or cat food) with regular vitamin/mineral supplements. Its close relative, the Jaboty or Yellow-legged Tortoise, *G. denticulata*, requires similar care.

Common Snapping Turtle
Chelydra serpentina

Members of the family Chelydridae, snapping turtles need to be handled with extreme caution: their razor sharp jaws are capable of giving nasty bites. The maximum length is about 35 cm (14 in). It is a uniform dull brown in color and has a relatively small shell and a thick, crested, tapered tail. It has a wide range, being found throughout eastern and central USA and south as far as Ecuador. It is an easy aquarium pet, though it tends to remain vicious. It will do well on a diet of fish and meat, including dog and cat food. The Alligator Snapping Turtle (*Macroclemys temminckii*) is one of the largest freshwater turtles, with a carapace length reaching 60 cm (24 in) or more and weighing more than 100 kilograms (220 pounds). Because it is so large and has become uncommon recently, it probably is best that amateurs avoid it; its bite can definitely crush a finger.

Eastern Mud Turtle
Kinosternon subrubrum

Mud and musk turtles belong to the family Kinosternidae and are capable of producing a foul smell from glands under the edge of the carapace when they are alarmed. Happily, most of them lose this habit after a short time in captivity. The Eastern Mud Turtle is found over much of the southeastern USA. The shell is uniformly olive to dark brown and the plastron is double-hinged. The head is brownish striped, spotted, or mottled with yellow. Keep this turtle in a medium-sized aquaterrarium with more filtered water area than land. Maintain a water temperature of about 24°C (75°F) and provide a basking spot on the land area. Feed on a selection of animal material, living and dead, especially fish.

Stinkpot or Common Musk Turtle
Sternotherus odoratus

In spite of its rather unglamorous name, the Stinkpot soon loses its unpleasant habit and becomes a well-behaved pet. One specimen lived in captivity for 55 years! It is found throughout the

Easy to acquire but sometimes nasty, the Common Snapping Turtle, *Chelydra serpentina*, is a voracious eater. Photo by R. T. Zappalorti.

eastern USA, where it lives in a variety of ponds, rivers and lakes. Rarely more than 15 cm (6 in) in length, the Stinkpot can be quite a pugnacious little beast when first captured, so watch for its vice-like jaws

The Mississippi Mud Turtle, *Kinosternon subrubrum hippocrepis*, and most of its genus-mates make adequate pets and are quite abundant throughout their range. Photo by R. D. Bartlett.

with which it can reach back as far as its hind legs! The smoothly rounded carapace is normally yellowish buff to olive in color, and there are two white stripes on either side of its head. Care for the Stinkpot as you would for the Eastern Mud Turtle.

CHELONIA

Red-eared Slider
Pseudemys scripta elegans

Belonging to the family Emydidae, this is probably one of the most familiar of all freshwater turtles and a very popular pet in several countries. In the past, wild hatchling specimens were collected in their millions in the southeastern parts of the USA and exported to many parts of the world as well as other parts of the USA for the pet trade. Fortunately the trade has now been regulated, and captive-bred or at least hatched stock from turtle farms is now available. Because of health laws, in the USA and Canada it is illegal to sell any Red-ears (and other water turtles) less than 100 mm (4 in) shell length. The Red-eared slider is possibly the most attractive of the four subspecies of *Pseudemys scripta* that occur in the southern USA in densely vegetated, slow-moving rivers, streams, ponds, and swamps. (Another dozen or more subspecies occur further south, reaching to central eastern South America.) Reaching a maximum length of 30 cm (12 in) or so, its shell is attractively mottled in dark green and black, becoming more somber as the turtle matures. The major distinguishing feature of the subspecies is the bright red patch on each side of the head.

Hatchling Red-ears require a great deal of care to raise satisfactorily. A hygienically kept aqua-terrarium and an adequate and varied diet are essential. Young turtles are particularly susceptible to mineral deficiencies, so a good vitamin/mineral supplement should be worked into the food. Provide extra calcium in the form of crushed egg shells or cuttlebone.

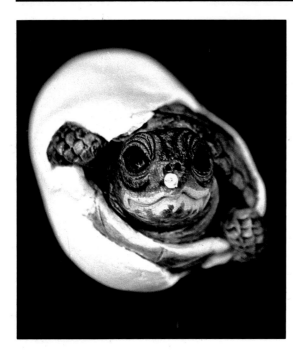

Both Pages: The Eastern Box Turtles, *Terrapene carolina*, make hardy, long-lived pets, but are best kept in outdoor enclosures. The one shown on the left, incidentally, is a ceramic model. Photos by Dr. Herbert R. Axelrod and L. Quinn.

Painted Turtle
Chrysemys picta

This is a small but very attractive species that reaches a total length of about 19 cm (7.5 in). The dark brown carapace is marked with bright red, especially within the marginal scutes. The limbs are also marked with red, and the head and neck are additionally marked with yellow stripes and patches. The underside is mainly creamy yellow. It occurs in much of eastern and northern USA and adjacent Canada. Its requirements are similar to those of the Red-ear though adults are more carnivorous.

BOX TURTLES

These are turtles that are almost totally terrestrial but commonly live in marshy or damp woodland areas. They belong to the family Emydidae along with the more common aquatic turtles, not the Testudinidae with the true tortoises. Two species are available on a regular basis and make good pets for the terraria of more advanced hobbyists.

Eastern Box Turtle
Terrapene carolina

This species occurs in the eastern, southern, and central USA, where it usually inhabits woodlands and the margins of fields. Color and pattern vary tremendously, both individually and according to range. They are relatively small but high-domed turtles, with the carapace length rarely exceeding 17 cm (7 in). Most are attractively colored with yellow, orange, and dark brown stripes and blotches on the head, limbs, and shell. They are omnivorous, feeding on a variety of insects, crustaceans, molluscs, and some carrion, as well as fruits, leaves, and some fungi. Captive specimens can have their diet supplemented with a little lean meat to which some vitamin/mineral supplement has been added. In captivity, most specimens will greedily take sweet succulent fruits such as strawberries, cherries, plums, and so on. It is best kept in an outdoor enclosure and allowed to hibernate naturally. The closely related Western Box Turtle, *T. ornata*, from the central and southwestern USA, requires similar but somewhat drier conditions.

ASIAN BOX TURTLES
Cuora species

Perhaps a little more aquatic than the American box turtles are the 8 to 10 species of Asian box turtles in the genus *Cuora*. Probably the best known and most attractive pet in the group is the Ambon Box Turtle (*C. amboinensis*). Reaching a total length of 20 cm (8 in), this charming species has a highly domed carapace that is dark greenish brown. There are vivid yellow stripes on each side of the head. Often found in and around rice paddies in Southeast Asia, they are said to be almost totally herbivorous, but captive specimens will take lean meat and sometimes mealworms. They require a large aqua-terrarium heated to 30°C (86°F), cooled a little

Within the last decade or so, soft-shell turtles have become more and more popular in spite of their irascibility. Photo by J. Mehrtens.

at night. Feed on a mixture of fruits and greens as well as lean meat and insects. A weekly vitamin/mineral supplement will not go amiss.

SOFT-SHELLED TURTLES
Though sometimes vicious and snappy, some of the "soft-shells" have their fans. Almost totally aquatic, most soft-shells only leave the water at egg-laying time. They are quite different from other turtles in having the usual horny scutes of the outer shell replaced by a

Many adult soft-shells, like this Florida soft-shell, *Trionyx (Apalone) ferox*, can deliver a nasty bite and are capable of reaching back at least half the length of their shells. Photo by K. T. Nemuras.

CHELONIA

soft, leathery skin that sometimes has unusual colors and patterns. Even the bony ribs that comprise the shell of other turtles are reduced in extent so the back and sides of the carapace are very flexible. The snout of a soft-shell is elongated, so the reptile can lay submerged in its muddy environment and be able to breathe without surfacing. All species of soft-shell are carnivorous, feeding on a variety of fishes, other aquatic creatures, and carrion. Soft-shells must be handled with care as they are capable of giving a nasty bite with their razor-sharp jaws.

Florida Soft-shell
Trionyx ferox

Of all the soft-shelled turtle species, the Florida Soft-shell seems to be the one most widely kept as a pet. Occurring in Florida and adjacent states, it grows to about 35 cm (14 in) in length and can reach over 60 cm (24 in). The color varies across a range of dark brown and green mottlings. In captivity, these soft-shells require a large aquarium with a thick sandy substrate (without rough or sharp edges) in which they can hide and water shallow enough for the animals to reach the surface with the tips of their snouts without emerging from their refuges. They may be fed on a variety of animal matter. Strict hygiene and water filtration are essential.

SUBORDER PLEURODIRA

All of the turtles discussed above belong to the suborder Cryptodira, the "straight-necked" turtles, so-called because they pull their necks into their shells in the shape of a vertical "S". The side-necked turtles of the suborder Pleurodira form about one-fifth of the world's chelonian species. These pull their necks into their shells in the shape of a horizontal "S".

Helmeted Turtle
Pelomedusa subrufa

Native to the water bodies of much of Africa south of the Sahara, the Helmeted Turtle was once commonly found in pet shops of Europe. Averaging 15-18 cm (6-7 in) in length, it is rather plain dark brown in color but makes an interesting inmate for the aqua-terrarium. Its plastron is hinged and it is able to withdraw completely into its shell. It must be kept at a water temperature of 26°C (79°F) and should have a warmer basking area on land. It will take most animal-based foods.

Australian Snake-necked Turtle
Chelodina longicollis

Captive-bred specimens are occasionally available. As its name implies, it has a very long neck that is just as long as its shell, which reaches 25 cm (10 in). It is a fairly plain dark brown or gray in color. An easy and entertaining inmate for the aqua-terrarium with water temperatures around 25°C (77°F), it will become very tame and take food from the fingers. It is found in suitable bodies of water throughout eastern Australia and feeds on a variety of animal matter.

Matamata
Chelus fimbriatus

This is one of the most bizarre of all turtles. It inhabits sluggish and stagnant waters in tropical South America, lurking in the bottom mud and vegetation, perfectly camouflaged by strange warts and tubercles on its head and limbs. The carapace scutes are also unusually pyramid-shaped to further enhance its strange appearance. The mouth is huge. If a fish or other aquatic creature gets close enough, the jaws are opened so quickly as to cause a strong water current that sweeps the

Unusual and attractive, the Matamata, *Chelus fimbriatus*, ventures onto land only to lay its eggs. Photo by Dr. Herbert R. Axelrod.

victim into its gullet! Almost totally aquatic, the Matamata only comes onto dry land to lay its eggs. Growing to a maximum of 45 cm (18 in), it requires a large aqua-terrarium with a small land area in case breeding is to be attempted. Hold the water temperature to around 26°C (79°F). Feed on fishes or large aquatic invertebrates.

The Crocodilians Order Crocodilia

The crocodilians comprise the crocodiles, alligators, caimans, and the gharial. They are all amphibious carnivores feeding mainly but not exclusively on fish—some of the larger examples are known to take large mammals, including man, should the occasion arise. The Nile Crocodile, *Crocodylus niloticus,* of Africa and the Estuarine Crocodile, *C. porosus*, of Asia and northern Australia are particularly noted for their occasional man-eating activities. The crocodilian body is covered with leathery cutaneous scales covering bony plates on the back and sometimes on the belly. Crocodilians are largely aquatic, inhabiting the rivers and lakes of warmer parts of the world. The eyes, valved ears, and valved nostrils are set high on the head so that the reptile can lie almost completely submerged but keep its major sense organs above water. The powerful tail is laterally flattened to aid in swimming, and the limbs with their partially webbed feet are used for steering. Crocodilians comprise three subfamilies: Crocodilinae with three genera and 13 species of true crocodiles; Alligatorinae with four genera and 7 species of alligators and caimans; and Gavialinae with a single genus and species, the long-snouted Gavial or Gharial from the Ganges River and adjacent areas of India and Bangladesh. All species are at least minimally protected

Although certainly part of the reptile community, crocodilians like this *Alligator mississippiensis* are not recommended as pets. Photo by R. T. Zappalorti.

by law, with imports today few and far between.

Most crocodilians make poor pets for the beginning hobbyist. Although hatchling crocodilians may look cute and can be easily kept as pets, they soon outgrow their accommodations and can become a dangerous menace in the home. They are therefore not recommended for the home terrarium keeper. At one time there was an excess of home-reared crocodilians offered to zoos on a regular basis, often without takers. Due to the demand for crocodile leather, many crocodilians became severely endangered at one stage, but this situation has now been alleviated to some extent by protecting wild populations and farming others to supply the demand.

Tuatara Order Rhynchocephalia

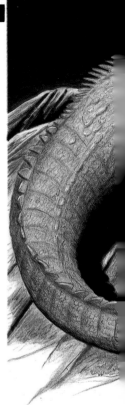

Shown here is a beautiful rendering by John R. Quinn of the rare and intriguing Tuatara, *Sphenodon punctatus*.

Although the Tuatara is never likely to be kept by the average terrarium keeper, it should, for interest's sake, be included in any book about reptiles. Looking superficially like a lizard, the Tuatara, *Sphenodon punctatus* is the only living member of its genus, family (Sphenodontidae), and order. (Recently a possibly undescribed second species has been recognized.) The Tuatara is the sole survivor of a very ancient group of reptiles and differs from the lizards mainly in aspects of its skull structure, including having the upper jaw formed into a "beak" at the tip. It is now found only in New Zealand on about 30 islands off the coast. Growing to a maximum of 60 cm (24 in), the Tuatara is a uniform olive-green to gray. It has short, well-developed limbs, a spiny

crest, and a substantial tail. It lives in burrows on its cold, humid islands, often sharing these with sea birds. It feeds mainly on invertebrates but has also been known to take small lizards and the chicks of sea birds. The Tuatara is probably one of the slowest growing and longest living of the reptiles. They do not reach sexual maturity until they are 20 years old and are thought to live for over 100 years.

Squamata, The Snakes and Lizards

To this, the largest existing reptilian order, belong the several thousand species of lizards and snakes, plus the obscure and little-understood amphisbaenians, the worm-lizards. The snakes probably evolved from lizard ancestors, judging from skull and soft tissue structure, becoming specialized for lives without limbs. Tuataras are closely allied to the squamates, and today some scientists feel they are not distinct enough to be full orders.

Lizards
Suborder Lacertilia

There are approximately 3000 living species of lizards in about 16 families within the suborder Lacertilia, sharing the order Squamata with the snakes. The typical lizard is an agile, four-limbed, scaly creature. However, there are many species far from typical, some having reduced limbs, only two limbs, or no limbs at all. Legless lizards can usually be distinguished from snakes by the presence of movable eyelids and/or external ear openings, neither of which is possessed by any snake.

Many lizard species make excellent terrarium pets provided they receive adequate care. Found throughout the warmer parts of the globe, some lizards may be difficult to supply with the specialized environmental conditions and/or diet needed to maintain them in good health, so a great deal of thought should go into the selection of species.

THE GECKOS

FAMILY GEKKONIDAE

The geckos comprise one of the larger lizard families with about 600 species. Relatively small lizards, the largest common species reaches just 15 in (37 cm) (*Gekko smithi* from Southeast Asia). Geckos are noted for their ability to vocalize and their remarkable climbing abilities, many species having special gripping mechanisms on the soles of their feet and toes.

Western Banded Gecko
Coleonyx variegatus

This is a small gecko reaching a maximum of 12 cm (5 in). Being in the subfamily Eublepharinae, it and other members of the same subfamily depart from some of the typical gecko characteristics. The digits are without adhesive pads (lamellae) and the eyelids are movable (fused in other subfamilies). The delicate-looking skin is yellowish white marked with chestnut-brown bands that extend onto the robust tail. It occurs in the western parts of the USA from California through Arizona and northern Mexico. It is a ground-dwelling, crepuscular species requiring a small, dry, heated terrarium with gravel and rocks. The average daytime temperature should be 22-28°C (72-82°F) reduced by a few degrees at night and for a winter rest period of 2-3 months. Feed on small invertebrates.

Leopard Gecko
Eublepharus macularius
This popular terrarium pet is a robust, large-headed species with movable eyelids. Like the Banded Gecko, it is not a typical climbing gecko and does not possess adhesive lamellae on the feet. Reaching 20 cm (8 in), its granular skin is grayish buff, patterned with the leopard-like dark spots and blotches that give it its common name. Juvenile specimens are quite differently marked from the adults, having stripes and

SQUAMATA

Two attractive geckos. Left: The Desert Banded Gecko, *Coleonyx variegatus variegatus*, photo by B. Kahl. Right: The Leopard Gecko, *Eublepharus macularius*, photo by Dr. Guido Dingerkus.

bands rather than spots. It requires housing similar to that for the Banded Gecko.

Tokay Gecko
Gekko gecko

Probably the most popular of all pet geckos (though the Leopard is catching up fast), this large (32 cm, 13 in) species is named after its two-syllable call, said to resemble a sharp "to-keh." Being in the subfamily Gekkoninae, it possesses the typical gecko characteristics, including fused, transparent eyelids, vertical pupils, and well-developed adhesive lamellae on the digits that enable it to cling to almost any surface, even upside-down. Though its teeth are relatively small and unlikely to break the human skin, an angry Tokay Gecko will bite with a painful, vice-like grip and will hang on like a bulldog until you can pry its jaws open. It should be handled with respect! Fairly colorful, the Tokay has a robust body and large head. The ground color of bluish gray is covered with numerous pink and sky-blue spots. Hailing from the warmer, wooded parts of Southeast Asia, this species requires a relatively large, tall terrarium with hollow logs and plants to provide cover. High humidity and temperatures in the range 26-32°C (79-90°F) should be provided. Feed on a variety of large invertebrates and even small mice. It will drink from a small water container or lick droplets sprayed onto leaves and other surfaces.

Chit-Chat
Hemidactylus frenatus

The common house gecko of Southeast Asia, the much loved Chit-Chat (named after its call) has spread with the help of human transport to many parts of the world and is now a common animal in coastal situations throughout the tropics. In its range it may be encountered everywhere from hollow trees in remote areas to inside city apartments, where it is welcomed for its insect-catching abilities. The geckos emerge from their daytime hiding places at dusk and congregate around light sources to hunt the

Perhaps the most popular of all geckos in today's hobby is the Tokay Gecko, *Gekko gecko*. Photo by H. Hansen.

insects attracted there. They are patterned in various browns that may darken or lighten according to mood and/or light intensity. In captivity provide Chit-Chats with a tall terrarium with cork bark or something similar attached to the inside walls, providing excellent daytime refuges. One or two potted plants will enhance the humidity. Mist spray daily. The temperature should be in the range of 24-32°C (75-90°F). Feed on small insects, especially flies and moths.

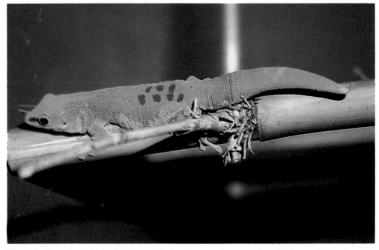

The Madagascar Day Gecko, *Phelsuma madagascariensis*, is active primarily during the daytime and requires a high-calcium diet. Photo by Elaine Radford.

The closely related Turkish Gecko, *H. turcicus*, has also become introduced to foreign areas, including the coastal ares of Texas and Louisiana as well as large areas of Florida. It requires similar care to the Chit-Chat.

DAY GECKOS
Phelsuma species

These fascinating geckos are popular due to their (usually) bright coloration and the fact that they are active during the day. Most are bright green marked with combinations of reds, blues, and/or yellows. Being diurnal, the pupils of the eyes are round, compared to the vertical shape in the typical geckos. There are several species found on the various islands and coastal areas of the western Indian Ocean, including

SQUAMATA

Madagascar, Mauritius, Seychelles, Comoros, Zanzibar, etc. Though most are protected, many are bred on a regular basis in captivity so are often available. Arboreal in the wild, they require well-planted terraria and temperatures in the range of 25-30°C (77-86°F). They feed on a variety of insects and nectar. In captivity their diet may be supplemented with honey (50/50 water and honey) and/or a sugar lump to which a drop of fluid vitamin/mineral solution has been added. Provide digestible calcium (crushed cuttlebone or egg shell) in a separate little cup because these lizards have very high calcium requirements.

IGUANAS

FAMILY IGUANIDAE

Containing some 700 species in about 60 genera, the Iguanidae forms the largest lizard family. Diverse

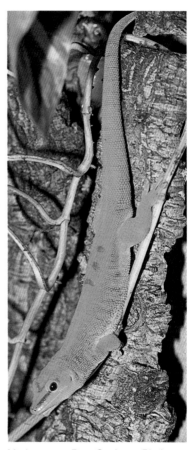

Madagascar Day Geckos, *Phelsuma madagascariensis*, also need a high level of humidity. Photo by H. Zimmermann.

in form and habit (some herpetologists have suggested that they form as many as eight distinct families), iguanids may be terrestrial, arboreal, semi-aquatic or even semi-marine (*Amblyrhynchus*, the Galapagos Marine Iguana, which feeds principally on seaweed). Iguanid headquarters are the Americas from southern Canada almost to the tip of South America, but a few species exist on Fiji and Madagascar. Most species possess a long, whip-like tail, and many have a spiny dorsal crest. Some are almost totally herbivorous, though most of the smaller species feed mainly on animal matter.

Green Anole or American Chameleon
Anolis carolinensis

This species gets its alternative common name from the fact that, like the chameleons of the Old World, it can change its body color dramatically. However, the similarity ends there and the two groups are otherwise only distantly related. The genus *Anolis* is a truly large group of some 400 species ranging over virtually all of the Americas and usually called anoles by herpetologists. Perhaps it is time to retire the misleading name "American chameleon" and replace it with the more distinctive "anole." Reaching a total length of 20 cm (8 in), this anole has a long, slender body and tail, a narrow head, and a somewhat pointed snout. Males are larger than females and possess extendible dewlaps that they show during sexual or territorial activity, exposing their bright pink or red color. The body color is green, gray, or brown and can be rapidly changed, depending on the mood, temperature, and health of the lizard.

Unusual for iguanids, many anoles (including the

Green) possess adhesive digital lamellae similar to those of geckos. This allows anoles to be extremely adept climbers. This species is found in the southeastern USA from North Carolina to eastern Texas, where it occurs in trees and shrubs, fences, and even house walls. It is a diurnal, sun-loving lizard that requires a tall, well-lit terrarium with plenty of climbing branches and foliage plants. Maintain the temperature at 23-26°C (74-79°F), reduced slightly at night. Local basking spots may have higher temperatures. Feed on a variety of small invertebrates and spray the foliage daily so that the lizards have water droplets

Although occasionally available in the herp hobby, Basilisks like this *Basiliscus plumifrons* are somewhat delicate and require much attention to detail. Photo by Alex Kerstitch.

to drink. The larger Knight Anole (*A. equestris*) of Cuba, introduced into Florida, requires larger housing but similar husbandry and will even take baby mice.

BASILISKS
Basiliscus species

Due to their bizarre appearance, basilisks are popular terrarium subjects. There are about five species, but the most spectacular must be the Green or Plumed Basilisk (*B. plumifrons*). Reaching a length of 65 cm (26 in), of which more than half is tail, it is a slender, long-limbed lizard possessing a double crest on the head, another some 5 cm (2 in) high along the back, and a third one along the tail. It is bright green in color with lighter patches merging into bluish green along the flanks and bright blue spots. It lives in Central America, where it occurs in trees, usually close to open water, into which it may leap if disturbed. Some basilisks can run so fast on their hind legs that they can move over the surface of water without sinking! Basilisks require a spacious terrarium with plants, climbing branches, and a spacious body of water. Maintain temperatures in the vicinity of 24°C (75°F) (night) to 30°C (86°F) (day) and provide good broad-spectrum lighting. Feed them on larger invertebrates, baby mice, and regular vitamin/mineral supplements. All basilisk species require similar care.

Desert Iguana
Dipsosaurus dorsalis

Occurring naturally in the southwestern USA and Mexico, this species has a relatively small head and a plump body. Growing to a maximum of 40 cm (16 in), it has a short crest running the length of the back and into the tail. It is usually light brown to cream with bands of dark brown and some yellow spots. When basking in the sun, the body color lightens dramatically. It is a diurnal lizard that can withstand high basking temperatures. During the day it climbs into bushes, but at night it hides in burrows. It should be provided with a spacious, dry terrarium with plenty of gravelly floor space and hiding places, plus a couple of climbing branches. Local basking spots can be as high as 40°C (104°F), but these should be switched off at night and the terrarium reduced to room temperature. Primarily herbivorous, the Desert Iguana will feed on a variety of vegetation and seems to be fond of pungent herbs in captivity; some insect food may also be taken.

Note the leash on this Common Iguana, *Iguana iguana*. Photo by David R. Moenich.

Green or Common Iguana
Iguana iguana

This is probably the most loved and popular of all lizard pets, due to its gentle manner, bizarre appearance, and relatively easy care. Native to southern Mexico, Central America, and northern and central South America, its status is threatened in many areas due to loss of habitat (forest

Male Rainbow Lizards, *Agama agama*, can be distinguished from the females by their brighter, more colorful heads. Photo by Otto Klee.

situation somewhat. It is a robust lizard with an elongate body and a spiny crest extending along its back and into the long whip-like tail. The male has a large leathery dewlap that is extended during sexual or territorial display. The ground color is gray-green with darker bands, but sometimes the animal can be bright green or even brownish or rust colored. There is much variation in color pattern among individuals. The juveniles are bright green up to about twelve months of age, when the browner adult colors usually develop gradually. In the wild it is primarily arboreal and usually found in the vicinity of water, where it will take cover if disturbed. It is an accomplished swimmer. Females often migrate to communal nesting areas where the eggs are buried about 30 cm (12 in) deep in sandy substrate.

Reaching a total length of

clearance) and collection for food and the pet trade. Attempts at protecting the reptile in many countries may have alleviated the

160 cm (63 in) a pair of adults require a spacious terrarium (minimum 200 cm long, 100 cm high, and 100 cm deep, 78 x 39 x 39 in). Strong climbing branches and a voluminous water vessel are required. Maintain at 25-30°C (77-86°F), reduced by a few degrees at night. Broad-spectrum lighting is essential for good health. Feed on a variety of fruit and vegetation, though some insects and other animal matter may be taken. Hatchlings are mainly insectivorous, becoming omnivorous and finally largely herbivorous as they mature.

THE AGAMIDS

FAMILY AGAMIDAE

With over 3900 species in 35 genera, the agamids comprise the Old World equivalent of the Iguanidae, with some species from both families showing some quite remarkable similarities in habits and appearance. The true chameleons are very closely related to the agamids, and recently it has been suggested that they form only a single family that would be called the Chamaeleonidae. It is doubtful that this suggestion will be accepted for quite a while. The agamids are found in a great range of habitats within the geographical regions of Africa, Asia, and Australia.

Rainbow Lizard
Agama agama

This lizard is a common sight in and around many villages in central and western Africa, the brightly colored males head-bobbing and skirmishing as they argue over territory. Growing to about 40 cm (16 in) in length, it has a robust body, broad triangular head, and a wide jaw. The limbs are well-developed. The whip-like tail makes up about half of the total length. It has a small crest on the nape of the neck. The usual color is a fairly uniform reddish brown, but displaying males take on a wide variety of startling colors, including orange heads, bright blue patches on back and limbs, and yellow patches on the flanks and tail. Basking females also brighten up somewhat, but not as conspicuously as males. The Rainbow Lizard requires a relatively large terrarium with climbing branches and basking rocks. Basking areas may reach 45°C (113°F), but cooler areas should be available; the temperature should be reduced dramatically at night. Provide with a shallow container of drinking water and feed on a wide range of invertebrates and, occasionally, baby mice.

Bearded Lizard
Amphibolurus (or ***Pogonia***) ***barbatus***

Common in eastern and southeastern Australia, this species gets its name from the large pouch or "beard" that it inflates when alarmed; at the same time, it opens its mouth to expose the lemon-yellow interior. Growing to a total length of 45 cm (18 in), the basic color is mottled gray, darker beneath, but when sun basking or when excited, this becomes suffused with yellow and white blotches. In the wild, it inhabits a wide range of habitats from wet coastal woodland to semi-desert in the interior. It is diurnal and semi-arboreal, often ascending dead trees or posts in order to sun bask. Its habit of basking on roads has unfortunately led to large numbers of them being destroyed by traffic. It requires a roomy terrarium with infrared basking heat and facilities for climbing. The air temperature should be in the range 25-30°C (77-86°F) with local basking surfaces to 40°C (104°F). This and many other basking lizards will appreciate unfiltered natural sunlight in the summer if it

This Bearded Lizard, *Amphibolurus barbatus*, gets its name from the large "pouch" that inflates when it is angered. Photo by William B. Allen, Jr.

can be arranged. The temperature should be reduced at night, and a short winter rest period at reduced temperatures is recommended. Feed on a variety of invertebrates, baby mice, and regular vitamin/mineral supplements. May take some ripe fruit.

Asian Water Dragon
Physignathus cocincinus

Growing to a total length of 75 cm (30 in), this beautiful lizard is a popular vivarium subject. Superficially resembling the Green Iguana in shape, its body is bright green with touches of blue and pink around the throat, while the elongate, tapered tail is banded with brown. Native to Southeast Asia, it is diurnal, living in thickly forested areas, usually close to water, where it will take refuge if alarmed. It should be provided with a large aqua-terrarium with roughly equal areas of water and land. Sturdy basking branches reaching out over the water will make them feel at home. Air temperatures can reach 30°C (86°F), but reduce to about 23-25°C (74-77°F) at night. Keep the humidity high and provide broad-spectrum lighting with UV. Aquarium heaters in the water will be useful for maintaining the water temperature at about 28°C (83°F). Feed on larger invertebrates and small mice. Water Dragons usually will take lean raw meat and canned dog or cat food, and may take soft, ripe fruit. A regular vitamin/mineral supplement is essential. The closely related Eastern Water Dragon, *P. lesueuri* of eastern Australia is rarely available. This attractive species can be kept in a similar manner to *P. cocincinus*.

FAMILY CHAMAELEONIDAE

CHAMELEONS

This family comprises perhaps one of the most amazing groups of lizards. With their high-ridged, laterally flattened bodies covered with granular skin capable of sudden and astonishing color changes, and long, sticky, fly-catching tongue, they are indeed most fascinating. The eyelids are fused together to form a cone-shaped mound; the eyes are situated at the tip of the mound and are each capable of movement independent from the other. This enables the chameleon to view two completely different objects at the same time (prey and danger perhaps). Other features include the long prehensile tail and the opposable digits of the feet (permanently fused together in groups of two and three). There are over 100 species of

Facing page: Native only to Southeast Asia, the Asian Water Dragon, *Physignathus cocincinus*, makes a fair captive but needs many branches for climbing and may hide often. Photo by R. J. Koestler.

SQUAMATA

chameleon in at least four genera, all of them insectivorous and capable of catching insects on their long, sticky tongues. The species range from southern Europe to India and over the whole of Africa. They are particularly abundant on the island of Madagascar. Chameleons are not particularly easy captives and are in dire need of natural sunlight and fresh air. One way to compromise is to provide indoor/outdoor accommodations or a terrarium that can be moved outdoors as necessary. Ensure that the chameleons are safe and warm in inclement weather, however. As they are quite sluggish in general movement, chameleons may be placed outdoors in an isolated tree or shrub during suitably warm weather. They will thus benefit from the fresh air, sunshine, and the possibility of catching wild insects.

Panther Chameleon
Chamaeleo pardalis

This 40 cm (15 in) colorful lizard is from Madagascar and holds the distinction of being among the very first chameleons to be bred in captivity. It is long-lived and quite adaptable (most other chameleons die within weeks of importation), especially if you obtain only captive-bred individuals. Males have a bluish cast absent in females, a prominent crest at the back of the head, and a longish depressed snout. This species lays eggs. If you must keep a chameleon, try to obtain this species.

Jackson's Chameleon
Chamaeleo jacksoni

Only a few years ago, this species was regularly available on the pet market, but international legislation has now made it quite a scarcity in the terrarium. This is perhaps just as well, as over-collection was leading to it becoming

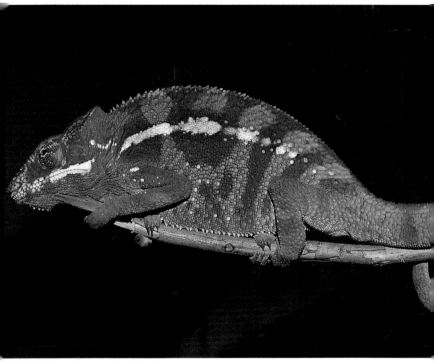

Many members of the chameleon genus *Chamaeleo* are now popping up in captivity. This one, known as the Panther Chameleon, *Chamaeleo pardalis*, was one of the first to be bred domestically. Photo by R. D. Bartlett.

endangered in the wild. Many captive specimens perished due to lack of adequate care. Sometimes called the Three-horned Chameleon, this species is remarkable in that the male possesses three relatively long, horn-like structures, an upward-curved one over each eye and another straight one on the tip of the snout. The female has only traces of these horns. The

It would be hard to imagine anyone taking offense to this cute, one-day-old Jackson's Chameleon, *Chameleo jacksoni*. Photo by J. Bridges.

color is usually bright green in fine weather and among green vegetation, but it changes rapidly depending on mood, position, and temperature to black, brown, blue-green, or gray, often with lighter whitish or yellowish spots and blotches. Found in eastern Africa, this species reaches 30 cm (12 in) in total length. It is a diurnal and arboreal species found at altitudes of 900 m (3000 ft) above sea level. It requires a large, airy, planted tall terrarium with an air temperature around 25°C (77°F), reduced to about 16°C (61°F) at night. Broad-spectrum lighting, good ventilation, and medium humidity are all important. Regular spraying of foliage will provide drinking water. Facilities to move the terrarium outside in suitable weather and allow unfiltered sunlight through a mesh lid should be arranged. A great variety of invertebrate foods should be given, and this should be dusted regularly with a powdered vitamin/mineral supplement.

FAMILY SCINCIDAE

SKINKS

Occurring on six continents, certain skink species are often the most familiar of lizards in some areas, living happily in cultivated gardens and performing a service by consuming insect pests. With over 600 species in 50 or more genera, the skinks form one of the largest lizard families. Most are elongate and circular in cross section with an indistinct neck. The scales are usually smooth, glossy, and overlapping. The majority have short but robust limbs, though in some species they are reduced or are partially or totally absent. Most skinks are terrestrial and many are specialist burrowers, but some can climb well and a few species are partially or totally arboreal. Most skinks are insectivorous, but some may take vegetable material. Many species are interesting and colorful and become tame and trusting in captivity.

Five-lined Skink
Eumeces fasciatus

This is a relatively slender species with a long, elegant tail and a small head. The young start life being almost black with white or yellow stripes and a bright blue tail. As they mature, the bright colors are replaced by a glossy brown, the male having a reddish tinge to the

Although not terribly common in today's herp hobby, skinks actually make perfectly adequate pets and can be very fascinating. Shown here is the beautiful and majestic Five-lined Skink, *Eumeces fasciatus*. Photo by R. T. Zappalorti.

This Eyed Skink, *Chalcides ocellatus*, would probably not make a very good captive since it likes to spend most of its time in hiding. Photo by Dr. Guido Dingerkus.

head, especially in the breeding season. Full adult size is about 23 cm (9 in). Its natural range is the eastern USA, where it is diurnal and terrestrial, inhabiting open woodland. It requires a semi-humid to humid woodland terrarium with a substrate of leaf-litter or something similar. Provide basking areas to 32°C (90°F) in summer, but reduce the cage temperature to room temperature at night. Allow to hibernate for 10-12 weeks at about 10°C (50°F) in winter. Feed on a variety of small invertebrates, provide a small water dish, and spray plants regularly.

Eyed Skink
Chalcides ocellatus

Growing to an overall length of 20 cm (8 in), this is a robust species of squat form, with short, powerful limbs and a shortish tail. The upper body color is bronze, with a sprinkling of black-edged, white "eye spots"; the underside is whitish. It occurs from southeastern Europe and the Mediterranean islands to Asia Minor and North Africa. Although active throughout the day, it is very secretive, hiding under ground litter, and rarely will be seen unless you are looking for it. In warm weather it is often active at night. It seems to prefer areas of stony scrub, vineyards, plantations, stone walls, and ruins. In captivity it requires a dry, semi-desert terrarium with a sand and gravel substrate. Provide flat stones or hollow logs in which the skinks will conceal themselves. Feed on a variety of invertebrates and ensure a supply of fresh water in a small shallow dish.

Eastern Blue-tongued Skink
Tiliqua scincoides

This is a large, docile species that makes an excellent terrarium "pet." Growing to 45 cm (18 in), it is a robust skink with a shortish tail and short but sturdy limbs. The color is light brown to silvery gray with darker cross-bands. It gets its common name from its bright blue, notched tongue that some people believe to be poisonous, but this, of course, is nonsense. Native to eastern and northern Australia, it is found in a range of habitats from dry sclerophyll to rain forest and open scrubland. Often it is found in suburban gardens, where it may raid the domestic dog or cat dish or eat fallen fruit in the orchard, but it also does a service by consuming lots of snails, the shells of which it crunches with obvious relish. It is mainly diurnal, hiding in hollow logs or under ground litter at night. It produces up to 25 live young and will breed readily in captivity. Give Blue-tongues a large terrarium with a shingle substrate, some leaf litter, and grass clumps. Hollow logs or rocks should be provided for hiding and basking. Provide basking temperatures to 35°C (95°F) and reduce the air temperature to around 20°C (68°F) at night. A short period of winter hibernation (6-8 weeks) is recommended. Medium humidity. Provide a large, heated water bath and feed on a variety of invertebrates (especially snails and earthworms), soft fruit, lean raw meat, and dog or cat food. The closely related Pink-tongued Skink, *Tiliqua gerrardi*, also from eastern Australia, requires similar care.

Solomons Giant Skink
Corucia zebrata

This is a somewhat atypical skink in that it is totally arboreal and has a prehensile tail. Recent breeding successes in zoological collections have made it more readily available to the home enthusiast, though still expensive. The head is short and broad. The robust body is light olive-green with faint brown bands. It occurs naturally only on the Solomon Islands and is arboreal and crepuscular to nocturnal, resting head-downward in thick vegetation or in tree hollows during the day. It requires a large tropical rainforest terrarium with adequate climbing facilities (robust branches and strong creeping plants). Daytime temperature may reach 30°C (86°F), but reduce to around 22-25°C (72-77°F) at night. Largely herbivorous, so feed on a variety of chopped fruit and vegetables, supplemented with a little chopped boiled egg and lean raw meat. Provide a regular vitamin/mineral supplement.

Facing Page: Note the bizarre tail on this Giant Sungazer, *Cordylus giganteus*. Photo by Elaine Redford.

FAMILY CORDYLIDAE

GIRDLED AND PLATED LIZARDS

This family is conveniently divided into two subfamilies: Cordylinae (girdled lizards) and Gerrhosaurinae (plated lizards). As a whole the family comprises some 10 genera and 40 species distributed widely over Africa south of the Sahara and on the island of Madagascar. The Cordylinae are typically spiny, with girdles of enlarged scales around the body and tail, while the Gerrhosaurinae have bony plates (osteoderms) beneath the scales. In most species the limbs are well-developed, but two genera (*Chamaesaura* and *Tetradactylus*) show varying degrees of limb reduction.

Giant Sungazer
Cordylus giganteus

The largest, or at least the most robust, member of the Cordylinae, this bizarre lizard grows to a maximum of 35 cm (14 in). It is a sturdily built species with spiny scales all over the back and tail. The head is broad and triangular, and there are large spines at the rear of the skull. It is mainly dark brown with darker blotches, the underside being grayish white with dark speckling. It occurs naturally in suitable habitats throughout southern Africa. A diurnal species that orients its head and body to get maximum warmth from the sun, this has led to its common name of Sungazer. It seems to prefer living in dry, rocky areas, often in fairly isolated populations in rocky outcrops. It is a popular pet lizard (when available) that soon becomes tame and trusting. In captivity it should be provided with a roomy, dry terrarium with a gravel substrate and piles of rocks. Local basking temperatures may reach 45°C (113°F). Broad-spectrum lighting is beneficial. Reduce the air temperature to about 20-25°C (68-77°F) at night. A winter rest period for a few weeks at reduced temperatures (10-15°C, 50-59°F) is also beneficial. It will feed on a variety of invertebrates, raw egg, lean minced meat, etc. It likes to soak in a shallow water dish.

Flat Rock Lizard
Platysaurus guttatus

Although also a member of the subfamily Cordylinae, first glances would find it difficult to relate this species to the sungazers. Small, relatively smooth, and colorful, only vestiges of the spiny girdles occur on the tail. The most remarkable aspect of *Platysaurus* (there are about 10 similar species in the genus), however, is the flatness of its body.

Dorsoventrally almost wafer thin, it can squeeze itself into the narrowest of rock crevices to escape from predators. Reaching a maximum of 25 cm (10 in), it is a fairly dull brown color in cool conditions, but after basking in the sun the male takes on brilliant green and blue hues with an orange tail. The female also colors up but not as dramatically as the male. Native to southern Africa, it lives in rocky conditions where there are plenty of cracks and crevices for refuge. It is a highly active, sun-loving species. It requires a large terrarium with artificial rock faces and controllable crevices as hiding places. It requires a daytime basking spot to 40°C (104°F), but with cooler spots also available. Reduce the temperature at night. Broad-spectrum lighting with UV is essential. Feed on small vertebrates and provide regular vitamin/mineral supplements. Water may be given in a very shallow dish or a rock hollow.

Sudan Plated Lizard
Gerrhosaurus major

A member of the subfamily Gerrhosaurinae, this is a well-known and popular terrarium subject. A docile species, it grows to a total length of about 40 cm (16 in) and has a robust body and short but powerful limbs. The tail is long and tapering. The squarish scales are arranged in rings around the body, and there is a prominent lateral fold in the skin along the flank. The body is a uniform reddish brown above, yellowish below. Its natural range is eastern and southeastern Africa. Diurnal, it lives in dry, sparsely vegetated, rocky areas. In captivity it requires a large semi-desert terrarium with a gravel substrate and a few large flat rocks. The daytime air temperature may reach 30°C (86°F), with warmer basking areas. Reduce to around 22°C (72°F) at night. Feed on a variety of invertebrates plus lean minced meat and dog or cat food; it may take some soft, ripe fruit. A shallow water bath is adequate for drinking and bathing purposes.

FAMILY LACERTIDAE

OLD WORLD TYPICAL LIZARDS

This family contains some 200 species in about 22 genera widely distributed throughout the favorable parts of Europe, Africa, and Asia. All are similar in having a "typical" lizard shape with an elongate body, well-developed limbs, and a long, tapering tail. The upper body scales are usually small and granular, while the belly scales are large and plate-like. The large head-scales are fused to the skull. Most species are diurnal, sun-loving, and extremely agile. They are the lizards most often associated with Mediterranean holiday resorts. Most species are insectivorous, but a few will supplement their diet with small vertebrates, fruit, or vegetation.

Canary Island Lizard
Gallotia galloti

Formerly included in the genus *Lacerta*, the three lacertid species of the Canary Islands have been found sufficiently unique to warrant their own genus. *G. galloti* is probably the best known species. Maximum length 35 cm (14 in). Males are dark brown to blackish with sky-blue spots on the cheeks and along the flanks. Females are lighter brown with dark stripes and lack the blue markings. They are native to the western Canary Islands, occurring on the grassy and rocky volcanic slopes, but sometimes also on walls around human habitations. In captivity they require a roomy terrarium

There shouldn't be much question as to why this beautiful little animal is known as the Green Lizard, *Lacerta viridis*. Photo by H. Hansen.

with gravel and rocks and perhaps a grass clump or two. Provide basking temperatures to 35°C (95°F), reduced to room temperature at night. Feed on a variety of invertebrates. They may take some soft fruit, especially banana. Provide a shallow water dish.

Green Lizard
Lacerta viridis

Growing to a maximum length of 40 cm (16 in), this is one of the best known European lizards. With an elegant tail twice the length of its head and body, this attractive species is bright green with a sprinkling of yellow and darker green

This common Wall Lizard, *Podarcis muralis*, is displaying a rather oddly-shaped tail, which is typical when the old one has been broken off.

spots. During the breeding season, the male's throat becomes bright blue. Females and juveniles are a somewhat more somber olive-green with a series of lighter and darker longitudinal stripes. It occurs in central and southern Europe and is an an active diurnal species found in open woodland, heathland, hedgerows, and plantations. It climbs into low trees and other vegetation to bask or search for prey. Soon becoming tame, it requires a large planted terrarium with medium humidity and facilities for climbing. The daytime air temperature should be maintained around 25°C (77°F), with hotter basking areas. Reduce to about 20°C (68°F) at night. A short hibernation period in the winter of 8-10 weeks at 8-12°C (47-54°F) is recommended. Feed on a variety of invertebrates, minced lean meat, dog or cat food, and maybe a little soft, ripe fruit. Provide a regular vitamin supplement and a large water dish. This species will do well in an outdoor reptiliary in suitable areas. The larger (60 cm, 24 in) and scarcer Eyed Lizard,

L. lepida, of Spain, southern France, N.W. Italy, and N.W. Africa requires similar care.

Wall Lizard
Podarcis muralis

A typical lacertid, the Wall Lizard comes in many subspecies and color varieties ranging through various shades of browns and greens and even black, with varying degrees of speckling and striping. Maximum length is 20 cm (8 in). Inhabiting central and southern Europe, it is an active diurnal species and the common lizard seen on walls, rocky slopes, ruins, and so on in many parts of Europe. It requires a relatively large, well-ventilated, semi-humid terrarium with piles of rocks on which it can climb. It should have access to basking temperatures of 35°C (95°F), but the air temperature should be reduced to 18-23°C (65-74°F) at night. A short period of winter hibernation is recommended. Provide a shallow water bath and feed on a variety of small

The Algerian Sand Racer, *Psammodromus algirus*, requires a lot of room in captivity but is less nervous than other lacertids and does fairly well. Photo by H. Hansen.

invertebrates. The closely related and possibly more attractive Ruins Lizard, *P. sicula*, of Italy and Yugoslavia requires similar care.

Algerian Sand Racer
Psammodromus algirus

Reaching a total length of 27.5 cm (11 in), this agile species has a tail more than double the length of the head and body. Its back is light to dark brown or olive, and there are two dark-edged yellow stripes down each side. There may also be a faint dark dorsal line. The belly is silvery white, sometimes with a greenish sheen. Males have large blue patches on the flanks just behind the front legs; these are much reduced in the females. Occurring in southern France, Iberia, and northeastern Africa, it prefers dry, rocky areas. It is an excellent climber, often ascending trees and shrubs in search of prey; it is also an adept burrower and will quickly disappear into loose soil in the presence of danger. In captivity it requires a roomy terrarium with sandy substrate, rocks, and climbing branches. Provide daytime air temperatures to 28°C (82°F), with higher temperatures at basking sites. Reduce the temperature at night. A short period of winter hibernation is recommended. Feed on a variety of small invertebrates. Provide regular vitamin/mineral supplements and a shallow dish of fresh water.

FAMILY TEIIDAE

TEGUS AND WHIPTAILS

This family is sometimes referred to as the New World equivalent of the Lacertidae, with which it shares many similarities. With some 200 species in about 40 genera, they are distributed from the northern USA through Central America to Argentina and Chile. Most

Although not often seen for sale in the herp hobby, the Jungle Runner, *Ameiva ameiva*, is, as you can see, extremely colorful. Photo by Paul Freed.

are typically "lizard-shaped," being slender, agile, and strong-limbed. A few, however, are almost limbless burrowers. The body scales are granular, those on the head being larger and plate-like. The tail is long and elegant.

Six-Lined Racerunner
Cnemidophorus sexlineatus

Reaching a total length of 30 cm (12 in), of which more than half is tail, this is a slim lizard with a pointed snout. The brownish body is marked with 6 or 7 evenly spaced yellow stripes. The male's throat is bluish or greenish, that of the female whitish; the underside is off-white. It occurs in the southeastern USA, reaching into Texas and New Mexico in the west. It seems to prefer sandy areas in open scrub, meadow land, and swamp margins. A terrestrial, diurnal, and sun-loving reptile that is extremely alert, active, and fast-running, it requires a large terrarium with plenty of floor area and a substrate of coarse sand with a few flat stones and grass clumps. Humidity should be low to medium and ventilation should be good. Maintain the air temperature around 28°C (82°F) with warmer basking sites. Reduce the temperature at night to 18-22°C (65-72°F). A short winter rest period at reduced temperature is recommended. Feed on a variety of small invertebrates plus regular vitamin/mineral supplement and provide a small dish of water.

Jungle Runner
Ameiva ameiva

This species has a robust body, powerful limbs, and a narrow head with a pointed snout. The tail is long and finely tapering, forming about half of the total length of 50 cm (20 in). The color is extremely variable, but usually shows varying amounts of bright green on the forequarters, running into brown at the rear. It occurs naturally in southern Central America and northern South America and has been accidentally introduced into Florida (Miami area). It is an extremely active, diurnal species with a very "nervous" disposition. Mainly terrestrial, it excavates burrows into which it retires at night or when danger looms. It is found mainly in open woodland, jungle clearings, and semi-scrub areas. It requires a relatively large terrarium with artificial, controllable burrows, gravel substrate, and a few prostrate plants. A medium to high humidity with air temperatures around 28°C (83°F) and basking areas to 37°C (99°F) should be provided. Reduce to not less than 20°C (68°F) at night. It will feed on a variety of invertebrates, some soft fruit, and a regular vitamin/mineral supplement. Provide a large water container.

Common Tegu
Tupinambis teguexin

The range of this large dark brown or black lizard

Be careful if you own one of these; the attractive Common Tegu, *Tupinambis teguexin*, has very sharp teeth, powerful claws, and doesn't take kindly to handling. Photo by K. T. Nemuras.

includes most of northern and central South America. The pattern varies considerably but usually includes spots or blotches of yellow, white or reddish on a darker ground color. It is found in a variety of habitats from tropical rain forest to fairly arid scrubland. They are diurnal and terrestrial lizards, retiring at night to their own constructed burrows. Tegus are active, powerful reptiles with sharp teeth and claws and must be handled with care. They require a very large, secure terrarium with a large water bath. Provide a heavy gravel substrate and a few hollow logs or artificial rocks, along with controllable "burrows" of plastic piping. Maintain medium to high humidity and a daytime temperature around 32°C (90°F), with warmer basking areas. Reduce to not less than 22°C (72°F) at night. They are fond of bathing, so should have a large pool or water dish. Feed on large invertebrates (snails especially), mice, chicks, minced lean meat, and raw egg. A regular vitamin/mineral supplement should be provided.

FAMILY ANGUIDAE

SLOW WORMS, GLASS SNAKES, AND ALLIGATOR LIZARDS

Anguid species are found in North and South America, Europe, and Asia. There are about 80 species in 3 subfamilies and a dozen or so genera. All species have an elongate body covered in overlapping scales supported by osteoderms, and many have a distinct fold separating the belly from the sides. The limbs are usually short, much reduced, or altogether absent. The relatively long tail can, in most species, be easily shed as a defense measure (autotomy), a replacement tail being grown in its place. The regenerated tail is always less elegant than the original.

Slow Worm
Anguis fragilis

A classical herpetological misnomer, the Slow or Blind Worm is neither particularly slow nor blind, and it is certainly not a worm. Growing to a total length of 35 cm (14 in), this limbless lizard is slimly built, with tight, overlapping scales. The movable eyelids and the ear openings are clearly visible—a sure sign that it's not a snake! The glossy scales are dull bronze in the adult, sometimes with a few pale blue spots. Juveniles are copper-colored, with a dark dorsal stripe. The underside is steel gray. *A. fragilis* is common throughout central Europe including Great Britain, but as it is very secretive in habits it is rarely seen unless searched for. It may often be found under pieces of flat metal, cardboard, or other rubbish lying on the ground. Mainly crepuscular in habit, it will bask in the early morning sun, especially in springtime. It inhabits hedgerows, heathland, open woodland, and gardens. An aquarium tank makes a good terrarium for two or three Slow Worms. They require a damp substrate (peat and sand mixture) with flat stones and grass clumps for hiding. A Slow Worm will appreciate summer basking temperatures to 28°C (83°F) but must have cooler refuges. Feed on small invertebrates, particularly small slugs and earthworms, and provide a small, shallow dish of drinking water. Allow to hibernate in the winter.

Sheltopusik
Ophisaurus apodus

This is a limbless lizard, often called a glass snake, that is serpentine in form and reaches a total length of 110 cm (44 in). It has been a popular reptilian pet for years in Europe, and specimens have been kept in captivity for upwards of 20 years. The eyes are well-developed and the lids are movable; the ear openings are conspicuous. A prominent lateral fold extends along the flank from

Sheltopusiks, *Ophisaurus apodus*, are not bad captives, but to find them being sold commercially is almost a hopeless venture. Photo by B. Kahl.

just behind the neck, ending in a tiny vestigial rear limb near the vent. The angular scales are arranged in rings around the body. The basic color is tan to bronze-brown, yellowish beneath. In the wild it occurs in southeastern Europe and Asia Minor. It inhabits open woodland, scrub, and rocky areas. It conceals itself in stone walls, under rocks, or burrows in loose soil.

Oviparous, the female coils around the developing eggs in a dark hollow, often beneath a large flat stone or piece of timber. Once the eggs have hatched she shows no further interest in her offspring. This species should be housed in a relatively large terrarium with a substrate of coarse sand and leaf-litter. Provide flat stones and a hollow log for hiding. Maintain daytime air and substrate temperature around 28°C (83°F), reducing it to about 20°C (68°F) at night. Humidity should be medium to low. A short winter rest period (8-10 weeks) at a temperature of 10-12°C (51-54°F) is recommended. Provide a shallow drinking vessel (in which the lizards may like to soak themselves). Feed on a variety of invertebrates, especially slugs and snails, raw egg, and lean minced meat; supplement with a regular vitamin/mineral concentrate.

Slender Glass Lizard
Ophisaurus attenuatus

Four species of glass lizards (*O. attenuatus, O. compressus, O. mimicus*, and *O. ventralis*) inhabit the eastern USA, the best-known being *O. attenuatus*. Reaching a total length of 106 cm (42 in), it is a stiff, limbless lizard with the typical skin-fold on the side of the body. It is a glossy light brown to buff with a black dorsal stripe and a pair of black stripes along each flank. At the anterior end the lateral stripes break up into black spots about the face and neck. Housing and care are very similar to that described for *O. apodus*. The tail is very fragile and likely to be shed in several pieces if the lizard is handled roughly, hence the common name "glass lizard."

Southern Alligator Lizard
Gerrhonotus multicarinatus

Reaching a total length of 30 cm (12 in), this is an elongate, slender lizard with a prehensile tail and short but well-developed limbs. There is a conspicuous lateral fold of skin between the fore and rear limbs. The color is variable but usually forms an attractive pattern of brown, black, and yellow arranged in bars, bands, and blotches. It occurs naturally in the western USA from southern

Washington to central and coastal California and adjacent Mexico. Inhabiting open woodland, grassland, and other dampish areas, it is primarily diurnal, frequently climbing into shrubs and other low vegetation in search of prey. Accommodations in captivity should consist of a large, tall, planted terrarium with facilities to climb. Ensure good ventilation and medium humidity and maintain daytime air temperatures around 25°C (77°F), with warmer basking sites. Reduce the temperature at night to 18-20°C (65-68°F). A short winter rest period at reduced temperatures is recommended. Feed on a variety of small invertebrates and provide a regular vitamin/mineral supplement. Provide a small drinking vessel and spray the terrarium plants 2-3 times per week. Other western USA members of the genus require similar husbandry, the temperature and humidity of course depending on the wild habitat.

FAMILY XENOSAURIDAE

CROCODILE LIZARDS

Containing only 2 genera and 2 species (one in Mexico and northern Central America, one in China), the robust bodies of these lizards are covered in scales supported by osteoderms and the sturdy limbs are well-developed. All give live birth and all are at least uncommon. The American genus, *Xenosaurus*, is virtually unavailable, while the Chinese *Shinisaurus* suddenly became available in numbers a few years ago only to virtually disappear within two or three years.

Crocodile Lizard
Shinisaurus crocodilurus

The common and specific names of this reptile are derived from the semi-

Native only to the westernmost parts of the U.S. and Mexico, the Southern Alligator Lizard, *Gerrhonotus multicarinatus*, has very powerful limbs and is quite quick. Photo by R. T. Zappalorti.

aquatic habit and the prominent double row of armored scales along the tail. Its color is mottled brown and white with reddish tones. Occurring in southeastern China, it inhabits damp woodland close to water. Basking on overhanging limbs, it usually takes to water if alarmed. Its maximum length is about 40 cm (16 in). It feeds on fish, tadpoles, and aquatic invertebrates. In captivity it requires a large planted aqua-terrarium with heated water. Maintain the water temperature around 15-20°C (59-68°F); this species must be kept cool. Feed on aquatic vertebrates and invertebrates.

FAMILY VARANIDAE

MONITOR LIZARDS

A small but spectacular lizard family containing a single genus and about 30 species confined to the sub-tropical and tropical parts of the Old World, including Australia. It includes the amazing Komodo Dragon, *Varanus komodoensis*, of Komodo Island and western Flores in Indonesia. With an adult length of over 3 m (10 ft) and a very heavy build, this is the largest living lizard species in the world. Other species range in size from 40 cm (16 in) to 2.4 m (8 ft). All have an elongate body, well-developed limbs, and a powerful, whip-like tail that is used in defense. (When handling monitors you must watch out for the bite, the claws, and the tail.) The head is usually long, with a pointed snout and powerful jaws. All species are carnivorous, feeding on a range of animals and carrion. Due to their large size and aggressive disposition, most monitor species are not recommended as a family pet, though some herpetologists have reared juveniles that have become remarkably tame, even on reaching adult size. Do not purchase a baby monitor lizard without weighing the consequences. These are large, sometimes dangerous lizards that need accordingly large and secure cages and a lots of care. They are not for beginners.

The Water Monitor, *Varanus salvator*, is among the largest of the world's lizards, some reaching a length of up to 8 feet! The specimen shown here is a baby. Photo by R. D. Bartlett.

Water Monitor
Varanus salvator

This is a large, robust lizard that reaches a total length of 240 cm (8 ft), much of it tail. It is dark gray to black with faint yellowish bands. In juveniles the markings are much brighter and more attractive, but the pattern fades as they mature. It occurs naturally in Southeast Asia, where it prefers wooded country close to water. It is an accomplished swimmer and climber. Give it a very large terrarium, preferably with a heated, drainable concrete pool and facilities for hosing down. Maintain the temperature around 26-30°C (79-86°F) throughout the year. Feed on mice, rats, chicks, whole eggs, whole fish, raw meat, and dog or cat food. Small pieces are preferred to whole animals, as monitors in nature are rather dainty eaters. With constant attention, specimens reared from hatchlings will become almost as tame and trusting as a dog. However, large wild-caught specimens rarely lose their natural aggressiveness and should be handled with extreme caution.

Savannah Monitor
Varanus exanthematicus

One of the smaller members of the genus, the Savannah Monitors are one of the more readily available species, and regular captive breeding may make them even more available. Reaching a length of about 90 cm (36 in) (larger in nature), it has a broader head and blunter snout than most other varanids. It is a robust reptile with granular scales. The basic colors are yellowish gray to brown with irregular lighter patches. Occurring naturally in savannah regions of eastern and southern Africa, it is a diurnal and terrestrial hunter and scavenger. It requires a large terrarium, preferably with a heated, drainable water bath. Care and food are as described for the Water Monitor.

FAMILY HELODERMATIDAE

BEADED LIZARDS

This family contains a single genus and two species, both of which are venomous, making them unique among lizards. Some of the teeth in the lower jaw bear grooves that conduct

Still very large, but not quite as much as the Water Monitor, is the Savannah Monitor, *Varanus exanthematicus*. These grow to about 36 inches. Photo by David R. Moenich.

venom from a number of labial glands. The prime purpose of the venom is to subdue prey. Bites from these lizards are seldom fatal to adult humans but are nonetheless painful and temporarily debilitating. Beaded lizards are definitely not recommended for amateur terrarium keepers, but they are well enough known and of enough popular interest to merit inclusion here.

The stout body is covered with small, granular scales supported by osteoderms. The limbs are short but well-developed, and the tail is shortish and plump. The Gila Monster, *Heloderma suspectum*, reaches a maximum length of 60 cm (24 in). The bead-like scales are mainly black, with a scattering of orange-red blotches. Its natural range is the southwestern USA into Mexico. It is a semi-nocturnal, desert-dwelling lizard that hides during the day in a burrow. Though generally sluggish, it is capable of striking rapidly at prey or aggressors. Gila Monsters should be kept in a very secure desert terrarium with a coarse sand/gravel substrate and controllable hiding places. Captive specimens like to immerse in water, and it seems to do them no harm. Allow daytime air temperature to reach 35°C (95°F), but reduce or turn off the heat at night. Feed on small mammals, birds, eggs, and raw minced meat, with a regular vitamin/mineral supplement. The Mexican Beaded Lizard, *H. horridum*, grows somewhat larger (80 cm, 32 in) and has whitish markings on the black scales. It requires similar husbandry to that described for *H. suspectum*.

Although popular in the herp hobby, the famed Gila Monster, *Heloderma suspectum*, is very unpredictable and somewhat expensive. Photo by Robert S. Simmons.

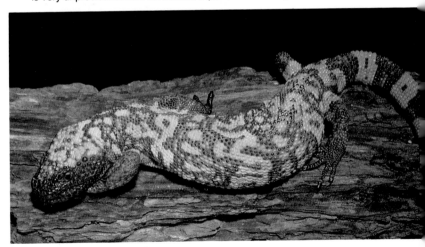

Suborder Amphisbaenia
Amphisbaenians

The amphisbaenians are the least known major group of reptiles and are rarely kept in the terrarium. However, this does not mean that they are uninteresting. The enterprising amateur herpetologist can perform his own research into the behavior of these "worm lizards." At one time they were included within the Lacertilia, but it is now generally accepted that they should have subordinal rank in parallel with the lizards and the snakes in the order Squamata. There are at least two families, Amphisbaenidae and Trogonophidae, which have 19 and 4 genera respectively. With the exception of the genus *Bipes*, which has well-developed forelimbs, all are limbless. The scales are arranged around the body in annuli, which resemble the rings of an earthworm. The word "amphisbaena" refers to the fabulous Greek serpent with a head at each end and a capability of moving equally adeptly in either direction. Although our worm lizards do not possess two heads, the blunt tail makes them look as though they might; some of the species even possess the ability of moving backwards when necessary—a useful attribute to any creature living in a narrow burrow. The eye is vestigial but visible as a faint dark spot. Amphisbaenians burrow and are often found in ant or termite nests. They occur in the warmer parts of North and South America, Africa, Europe, and southwestern Asia.

Florida Worm Lizard
Rhineura floridana

This example of an amphisbaenian reaches a maximum length of 40 cm (16 in) and is the only worm lizard species found in the USA. Occurring across central Florida, it resembles a large earthworm. It is opalescent pink in color and has a chisel-like snout as an aid to burrowing. Spending most of its time burrowing and feeding on subterranean invertebrates, it surfaces only after heavy rains and soil saturation. The best way to keep amphisbaenians is in narrow glass terraria. Two sheets of glass about 60 cm (2 ft) long set in a frame and separated by about 5 cm (2 in) would be about right for a couple of small worm lizards. Fill the terrarium with a mixture of sand and peat so that the animals can be viewed in their tunnels. The glass sides should be shielded from the light when you are not viewing the inmates. Feed on earthworms, termites, and similar invertebrates. Water can be administered via a concave stone or very shallow container placed at the soil surface. Be sure that the burrowing medium is slightly moist but never becomes waterlogged.

Snakes
Suborder Serpentes

It seems almost certain that the snakes evolved from burrowing or aquatic lizard-like ancestors, though the fossil evidence for this is extremely sparse. It is worth repeating here that the lizards and the snakes are closely enough related to be placed in a single order

SQUAMATA

(Squamata). There are about 2700 species of snakes worldwide in some 13 families. Typical features of the average snake are an elongate body and an absence of limbs (though a vestigial pelvic girdle is present in members of some of the more primitive families), immovable transparent eyelids, and absence of external ear openings. The skin is composed of overlapping scales that may be smooth and glossy, rough and mat, or keeled, depending on species. The teeth are sharp and recurved. In some families there are venom glands situated in the cheeks that supply venom to special fangs for the subduing of prey. The deeply forked tongue is used for detecting scents in conjunction with a pair of olfactory organs (Jacobson's organs) situated in the roof of the mouth.

Of all the reptilian groups, snakes bring out the greatest range of emotions in humans, and there are those who hate or fear them to a point of fanaticism, while others find them fascinating and interesting enough to want to keep them as pets. Venomous species are, in general, too dangerous to be kept in the home by amateurs and should be left to experienced specialists. However, there are many species of non-venomous snakes that make ideal pets or terrarium subjects. Some of these become tame and even seem to enjoy being handled—though other species have a naturally nervous disposition and are preferably admired by sight only, leaving handling to the minimum.

There follows a brief description of each of the snake families, though only examples of species regarded to be generally suitable for the home terrarium are described in greater detail.

FAMILY TYPHLOPIDAE

BLIND SNAKES

Primitive burrowing snakes, mostly small, species ranging from 15 cm (6 in) to 60 cm (2 ft), that are usually seen only after heavy rain or while turning soil. Because of their secretive habits, they and members of some of the other primitive burrowing families are not popular as home terrarium animals but are interesting study material for the potential herpetological researcher. The scales are uniform all round the body and the eyes are covered with enlarged scales. The mouth is extremely small and there are teeth only in the upper jaw. They feed on a variety of subterranean invertebrates, especially termites. There are about 240 species in at least 3 genera found in most of the warmer parts of the world (not in the USA or most of Europe), but they are particularly abundant in Asia and Africa.

FAMILY ANOMALEPIDAE

AMERICAN BLIND SNAKES

Members of this family of burrowers are similar in many respects to those of the Typhlopidae, but some possess a single tooth on the lower jaw. There are about 20 species in 4 genera that occur in Central and South America. Most are relatively small, reaching a maximum of 30 cm (12 in).

FAMILY LEPTOTYPHLOPIDAE

THREAD BLIND SNAKES

Though requiring revision, this family presently consists of about 40 species of burrowing snakes in two genera native to the Americas, Africa, and western Asia. These very slender and worm-like snakes are rarely more than 30 cm (12 in) in length. They feed extensively on termites and are often found in the nests of these insects.

FAMILY ANILIIDAE

CYLINDER SNAKES
Presently containing about 10 species in 3 genera, there is some dispute among taxonomists about the validity of this arrangement. Occurring in South America and Southeast Asia, they reach a maximum of 90 cm (36 in) in length.

Among the most primitive members of the order Serpentes are the blind snakes, family Typhlopidae. Photo of *Rhinotyphlops schinzi* by Paul Freed.

Red Cylinder Snake
Cylindrophis rufus

This is the best known of the cylinder snakes because it is found throughout Southeast Asia, often in the vicinity of human habitations. It is also the species most likely to turn up in the terrarium.

Unfortunately seldom seen in captivity, the blind snake *Liotyphlops albirostris* is, as you can see here, a very beautifully-formed animal. Photo by Paul Freed.

Growing to a maximum length of 75 cm (30 in), this species has glossy black scales over most of the body and is red beneath the tail. When alarmed, like all cylinder snakes it conceals its head among the body coils and exposes the red underside of the flattened tail. It requires a humid rain forest terrarium with facilities to burrow (clean leaf litter mixed with coarse sand). Feed on small vertebrates (frogs, lizards) or invertebrates, slugs, snails, etc. May be trained to take small dead fish. Maintain at about 30°C (86°F), reduced to about 25°C (77°F) at night.

FAMILY UROPELTIDAE

SHIELD-TAIL SNAKES

This unique group of about 8 genera and 40 to 50 species is restricted to India and Sri Lanka, where most species occur in rain forests and high mountains. They vary from only 10 cm (4 in) to over 50 cm (20 in) in length but usually are exceedingly slender. The scales are glossy, covered with minute parallel ridges that refract light into rainbow-like patterns. The tail often is covered by a specialized scale or group of scales that may have serrations or be greatly enlarged and project into a forked point. The tail scale, the shield, helps close the burrow. Few species are kept in captivity. Earthworms and other soil invertebrates are the major diet.

FAMILY ACROCHORDIDAE

WART, FILE, OR ELEPHANT'S TRUNK SNAKES

Unusual, almost totally aquatic snakes with one genus and 3 species. Ranging from India through Southeast Asia to northern Australia, they occur in fresh to brackish or even pure marine waters, but always near the coast. The skin, composed of small granular scales, hangs loosely on the body, providing the snakes with their common names. Unfortunately, the fine tanning quality of this skin makes these snakes, like the boas and pythons, vulnerable to the attentions of leather traders. Thousands of animals are killed each year (legally, it seems) for their hides, many of which are used for fancy boots and belts.

Javan Wart Snake
Acrochordus javanicus

Reaching a maximum length of 250 cm (8 ft 4 in) (males to only 145 cm, 5 ft), the granular skin of this species is pale creamy brown with darker marbling. The underside is whitish. Inhabiting coastal areas from Malaysia and through Indonesia to northern Australia, it is a sluggish reptile that usually lies among submerged vegetation, occasionally bringing its head to the surface for air. If an adult specimen is placed on dry land it seems almost incapable of moving its floppy body—though juveniles are semi-terrestrial. Feeding on fish and other aquatic animals, it forages by day or night. In captivity it requires a heated aquarium with water depth at least 30 cm (12 in) (provide an aqua-terrarium for juveniles). Aquarium heaters should maintain a water temperature around 28°C (82°F). Feed on live fishes (guppies, goldfish, trout, etc., of suitable size).

FAMILY BOIDAE

PYTHONS AND BOAS

The largest and some of the most spectacular snakes fall into this group and as such are very popular with the terrarium keeper. Most modern workers recognize 5 subfamilies, about 23 genera, and 80 odd species ranging

SQUAMATA

in size from dwarf burrowing species at 45 cm (18 in) to huge pythons in excess of 10 m (33 ft). Most species have a robust body and a relatively short tail. The dorsal scales are small and smooth, while the ventrals are broader than the dorsal scales but not as strap-like as the ventrals of colubrid snakes; they form a single row under the body and a double row below the tail. In most species the head is separated from the body by a relatively narrow neck. The mouth is large and has many recurved teeth situated in several rows on the palate as well as on the upper and lower jaws. Members of the family are renowned for their method of constricting prey by applying asphyxiating pressure with coils of the body (but not crushing the bones as is frequently thought).

Boa constrictor
Boa constrictor

One of the few snakes in which the scientific name is known and widely used by the general public, the Boa Constrictor is not the giant it is often made out to be when compared with the Green Anaconda or the Reticulated Python, for example. However, a large specimen of 3-4 meters (10-13) feet is a quite spectacular snake, though the average maximum size is usually somewhat less. It has a robust looking body, a broad triangular head, and a fairly narrow neck. There is great variation in color and pattern, but usually Boa Constrictors are buffy cream, boldly marked with dark brown blotches that become distinctly more reddish toward the tail. A member of the subfamily Boinae, the Boa Constrictor occurs in several subspecies ranging from Mexico to southern South America and a few southern West Indian islands. It occurs in a wide range of habitats throughout its range. Boa Constrictors require a large

Now in albino and red-tailed varieties, the Boa Constrictor, *Boa constrictor*, remains one of the most popular boids in the pet hobby. Photo by R. D. Bartlett.

terrarium with strong branches for climbing and a large, preferably heated, water bath. Air temperature should be maintained around 28°C (82°F) and reduced to about 20°C (68°F) at night. For the sub-tropical individuals a winter period of semi-hibernation, 4-6 weeks at 15-18°C (59-65°F), will stimulate reproduction. Feed on mice and chicks (juveniles), rats, small rabbits, guinea pigs, or chickens.

Rainbow Boa
Epicrates cenchria

Another member of the subfamily Boinae, the Rainbow Boa has several subspecies found from Costa Rica to northern Argentina. Although bred frequently in captivity, it is unfortunate that subspecific purity has been all but lost in captive populations. Maximum length is 2.5 m (8 ft), but usually smaller. The adult male is shorter and slimmer than the female. The common name arises from the beautiful rainbow bloom of the skin, especially in newly molted specimens.

Even those who don't have a particular preference for boids cannot deny the undying beauty of this Rainbow Boa, *Epicrates cenchria*. Photo by R. S. Simmons.

The color may be various shades of brown, uniform or with darker markings. Rainbows occur in a range of habitats and move just as well on the ground as among vegetation. It requires a medium-sized terrarium with facilities as described for *Boa constrictor*. Feed on mice, rats, and chicks. Other members of the same genus include Caribbean island species that are endangered in the wild, but captive breeding programs have made them frequently available to the hobbyist. These include the Cuban Boa, *E. angulifer*; the Puerto Rican Boa, *E. inornatus*; and the Haitian Boa, *E. striatus*.

Green Anaconda
Eunectes murinus

Perhaps no giant snake has fired imaginations so much as the Green Anaconda. Adventurers have returned from the depths of Amazon rain forests with stories of fantastic man-eating serpents 18 m (58 ft) or more in length. Although the Green Anaconda is certainly a big snake (second only in certified record length to the Reticulated Python), authentic and reliable records of total length are somewhat disappointing when compared to the dubious stories, and it seems that 9 m (29 ft) would be the maximum size. (Many people accept as valid a fairly well authenticated modern record of 11 m or 37 ft.) It is a thick snake, and a large specimen could have a mid-body diameter of 30 cm (12 in). The color is dull olive green marked with almost circular sooty brown blotches all down its back. Similar blotches along its flanks are centered with yellowish buff. The nostrils and eyes are set on top of its triangular head, indicating its fairly aquatic habit. Occurring in the tropical rain forests of northern South America, the Green

One of the largest snakes in the world, the Green Anaconda, *Eunectes murinus*, can grow to a length of well over 25 feet! Photo by Dr. Guido Dingerkus.

Anaconda is mainly nocturnal, living in or near watercourses, where it preys on fish, reptiles, water birds, and mammals. It will occasionally ascend into the branches of trees to sun bask. Juvenile specimens will tame reasonably well if handled frequently. As they grow to a large size, voluminous (preferably concrete and brick) accommodations are recommended. A large, deep (minimum 1 meter, 39 in), drainable concrete pool is required, and, as the snake will spend a lot of time submerged, some kind of

Only now being realized for its captive potential, the Rosy Boa, *Lichanura trivirgata gracia*, was formerly one of the most underrated of all pet snakes. Photo by F. J. Dodd.

water heating is recommended. Maintain the air temperature at about 30°C (86°F) during day, water temperature about 26°C (79°F). Reduce the air temperature at night. Feed on fish, mice or rats (juveniles); chickens, ducks, and rabbits (adults). The smaller Yellow Anaconda, *E. notaeus*, requires similar captive care.

Rosy Boa
Lichanura trivirgata

The Rosy Boa is a small but sturdily built species reaching a maximum of just 80 cm (32 in) in length. Its base color is bluish gray, and it has three longitudinal chocolate, reddish brown, or pinkish stripes that may be clearly defined with sharp edges or very broken and irregular with motley margins. Native to southern California and northern Mexico, the Rosy Boa is found in scrubland, open woodland, and rocky ravines. Though mainly terrestrial, it is an adept climber when the occasion arises and it feeds primarily on small mammals and birds. It should be housed in a medium-sized terrarium with facilities to climb and bathe. Maintain temperatures around 28°C (82°F), reduced at night. A short period of hibernation is recommended. Feed on mice and small chickens. *Charina bottae*, the Rubber Boa, is the only other boid native to the USA. It is delicate and requires unusually cool conditions; not suitable for captive care by amateurs.

Rough-tailed Sand Boa
Eryx conicus

A member of the subfamily Erycinae, this cylindrical snake has a very short tail that is the same shape as the head. Its dorsal scales are conical in shape and especially pointed on the tail, giving the snake its common name. Growing to about 80 cm (32 in) in length, it is perhaps the most attractive of the 10-12 species in the genus. The color is creamy buff with large, irregular reddish brown markings, these often bordered in darker brown. It is found in a variety of habitats from Pakistan through India to Sri Lanka. Like other sand boas, it spends most of its time buried under loose soil. Mainly nocturnal, it feeds on a variety of small mammals, birds, and reptiles. Provide a deep, dry substrate in a medium terrarium with one or two flat stones or logs. A small drinking vessel is adequate. Feed on mice or small chicks.

Green Tree Python
Chondropython viridis

Turning to the subfamily Pythoninae, the first species to be discussed is the Green Tree Python, sometimes referred to by its native Papuan name of "Jamomong." It is a laterally flattened snake with a broadly triangular head and is basically emerald green with a yellow or white vertical stripe. Juveniles may be brick red or yellow in color. Native to New Guinea, the Solomon Islands, and the northern tip of Cape York Peninsula in Australia, one notable aspect of this species is its very close morphological resemblance to the Emerald Tree Boa, *Corallus caninus*, of South America, an example of convergent evolution. It is a mainly nocturnal and strictly arboreal inhabitant of the rainforest. When at rest it drapes itself symmetrically over a branch. It requires a warm, humid, rainforest terrarium with daytime

Above: Although not terribly popular in today's herp hobby, the Rough-tailed Sand Boa, *Eryx conicus*, is an attractive species. Photo by C. Banks. *Opposite page:* As you can tell by this picture, the young of the Green Tree Python, *Chondropython viridis*, certainly are not restricted to one single color. Photo by Louis Porras.

temperatures to 30°C (86°F), reduced somewhat at night. Provide a pool of warm water and stout climbing branches. Feed on mice, rats, and birds.

Indian Python
Python molurus

The Indian Python is perhaps one of the most widely kept terrarium pythons due to its apparent docility once it settles. Even very large specimens remain very tame and gentle, provided they receive regular handling. (You should never become complacent about this, however—no snake can be completely trusted all the time.) Captive breeding is frequent, and most terrarium specimens today have been captive-bred. The dark phase or Burmese subspecies, *P. m. bivittatus*, is most commonly available. It is rich bronze-brown in color, marked with a network of broad cream and buff bands. The light phase subspecies, *P. m. molurus*, is similarly marked but in much lighter and duller

Note the unusual golden vertebral stripe on this variant Ball Python, *Python regius*. Photo by R. D. Bartlett.

shades. Only those with sufficient space should attempt to rear the larger boids. A fully grown female Indian Python can reach a massive 5 m (16.5 ft) or more in length, though males are generally at least one-third shorter and proportionately slimmer. This species is widespread throughout India and Sri Lanka, Bangladesh, Burma, Indo-China, Malaysia, and

Indonesia, where it tends to inhabit forested areas near water. It is becoming scarce in many parts of its range due to human predation (for food, folk medicine, and the leather trade) and loss of habitat due to land clearance. It requires a roomy, tropical terrarium with facilities to climb and bathe; a concrete or fiberglass drainable pool is most convenient. Feed on mice and chicks (juveniles), rats, chickens, and rabbits.

Royal or Ball Python
Python regius

The Royal Python is an ideal python for those with minimal space—if you can find one that will feed. Most specimens on the market are wild-caught and extremely shy. Most will feed only at night and in a secure cage. You must be willing to experiment to find a suitable food for your animal. Some will eat only live food, others only dead. Some will take only chicks, others white mice, still others only hamsters. Even color of the food may be a factor. Many Ball Pythons will never feed for their owners. Captive-bred specimens usually are much easier to feed and will take a more varied diet. The Royal Python is attractive in appearance and reaches a maximum of just 1.5 m (5 ft). It rarely attempts to bite, but nervous specimens will curl up into a ball with the head safely within the coils. This has given rise to its alternative name of "Ball Python." Attractively marked in chocolate brown and yellow-buff, it occurs naturally in the rain forests of West Africa where it is very secretive, often hiding among the roots of trees and low vegetation. Mainly nocturnal. It requires a medium-sized tropical terrarium with day temperatures to 30°C (86°F) and care otherwise similar to that described for *P. molurus*. It requires hiding places with a tight squeeze (hollow logs or rocks) before it feels secure enough to feed. Some keepers have recommended feeding Ball Pythons by putting them in a sack along with their dead food overnight.

African Rock Python
Python sebae

Often said to be slightly more aggressive than the Indian Python, which it superficially resembles, this species will nonetheless settle well into captivity. Reaching a maximum length of 5 m (16.5 ft), the male is somewhat shorter. This species inhabits many parts of Africa south of the Sahara. It is found chiefly in savannah, open woodland, and rocky escarpments rather than closed forests. Housing and care are as for *P. molurus*.

Reticulated Python
Python reticulatus

This is the longest of all living snakes, and specimens in excess of 10 m (33 ft) have been reliably reported, though the average adult size seems to be little more than 6 m (20 ft). In color it is light to dark brown with a reticulated pattern on the back of black, yellow, and buff. The flanks are marked with a double row of diamond or triangular white patches. It is an inhabitant of tropical Southeast Asia, where it usually is found in forests and fairly close to water. Rogue specimens occasionally enter villages to steal poultry and other domestic animals (including dogs, cats, young pigs, and so on). There are a few reports of this species attacking and attempting to swallow humans, usually children. Wild specimens are extremely aggressive and can give vicious lacerating bites with their numerous recurved teeth. Tamed from juvenile size and handled frequently, some may remain quite

The only snake longer than the Green Anaconda is the Reticulated Python, *Python reticulatus*, which has been known to attain a length exceeding 33 feet. Photo of mother with her eggs by William B. Allen, Jr.

docile, but there is no doubt that most Reticulated Pythons are dangerous animals. Housing should be as described for *P. molurus*, with a large, heated, and preferably drainable water bath.

FAMILY COLUBRIDAE

TYPICAL OR COLUBRID SNAKES

If any snake can be described as typical, then it belongs to this huge family. The Colubridae is extremely diverse, and various workers have divided it into as many as 14 subfamilies containing about 300 genera and almost 3000 species distributed throughout temperate, sub-tropical, and tropical parts of the world. Most colubrids are of medium size, slender, and agile, with broad ventral scales, but there are many variations. Many genera of colubrids make excellent pets. In some cases (such as *Elaphe* and *Lampropeltis*), captive breeding has been so successful over many generations that the species can be almost described as domesticated.

Western Whip Snake
Haemorrhois viridiflavus

Averages 150 cm (5 ft) maximum length but occasionally reaches 200 cm (6.5 ft). This and similar species often are treated in the genus *Coluber*, but many experts think that *Coluber* should be restricted to American species. *Haemorrhois* is available for the Eurasian racers or whip snakes and is being used more and more today. Native to western Europe from central France and Switzerland to Sicily, it is a rather slender snake with a greenish yellow ground color marked with black or dark green cross bars and blotches. It is mainly terrestrial though it can climb well among rocks and vegetation. It is mainly diurnal, is very fast, and hunts by sight. When newly captured it is extremely aggressive, striking and attempting to bite persistently. It requires a medium-sized, semi-humid terrarium with rocks and branches for concealment and climbing. A small water dish is adequate. Keep the daytime temperature at 26-30°C (79-86°F), reduced by a few degrees at night. Feed on frogs, mice, and chicks. A short hibernation period at reduced temperatures is recommended. Other closely

When this young Southern Black Racer, *Coluber constrictor priapus*, grows up, it will loose all of its patterning and become uniformly colored. Photo by R. D. Bartlett.

related Eurasian members of the genus include Dahl's Whip Snake, *H. najadum*; Horseshoe Whip Snake, *H. hippocrepis*; Balkan Whip Snake, *H. laurenti* (formerly *gemonensis*); and the Large Whip Snake, *H. jugularis*. All require similar husbandry.

Racer
Coluber constrictor

The only North American member of the genus (and the only species of the genus if the Eurasian racers and

Although not often seen in the pet trade outside of Europe, the Smooth Snake, *Coronella austriaca*, would probably make a fine captive. Photo by B. Kahl.

whip snakes are removed to another genus), it ranges from southern Canada to Guatemala and occurs in 11 poorly defined subspecies (commonly named Northern Black, Tan, Buttermilk, Blue, Eastern Yellowbelly, Blackmask, Brownchin, Western Yellowbelly, Mexican, Southern Black, and Everglades). A slender, agile, and fast-moving species reaching 200 cm (6.5 ft). Newly captured specimens share a bad temper with their European cousins. Colors range from uniformly black through bluish, greenish, or brown above, sometimes with yellowish or whitish spots; yellowish to grayish below. A diurnal species found in a wide range of habitats. Mainly terrestrial but climbs well. Feeds on a range of animals. In spite of its specific name it is not a constricting snake. Care as for *Haemorrhois viridiflavus*.

Smooth Snake
Coronella austriaca

A temperate species found over much of central Europe in suitable locations. Prefers dry sunny habitats, including heathland, hedgerows, open woodland, and disused railway tracks. Averaging 50 cm (20 in) in length (rarely reaching 80 cm, 32 in), it has variable coloring but is usually grayish to brownish with small dark spots and blotches along the body. It feeds largely on lizards in the wild but can be persuaded to take small mice in captivity. Keep it in a small, mildly heated terrarium with adequate cover and a small water dish. Hibernate for a few weeks in winter to induce reproduction.

Indigo Snake
Drymarchon corais

Reaching a length of 2.6 m (8 ft 8 in), the Indigo Snake is a spectacular species with several subspecies ranging from the southeastern USA through Central and South America as far as Argentina. The eastern subspecies, *D. c. couperi* is uniformly shiny blue-black with red tone on the head. It is much sought after by terrarium keepers and formerly was widespread in southeastern Georgia and Florida. It has, unfortunately, become endangered in the wild due to commercial collecting, gassing of gopher tortoise burrows (which the snakes like to use as refuges), and habitat destruction. It has been protected by law for some time, but captive-bred specimens are occasionally available at very high prices. The Texas Indigo Snake, *D.*

c. erebennus, has a brown and buff coloration that is less spectacular, but it still makes a charming terrarium subject, soon becoming tame and trusting. Other subspecies range in color from dull olive-green to brown, often differently colored at either end of the body. If alarmed or cornered, wild specimens will vibrate the tail tip rapidly, flatten the neck, and hiss loudly. Most Indigo Snakes prefer dry areas but usually near to permanent water. Diurnal and terrestrial, Indigo Snakes feed on a variety of vertebrates including other snakes, using the "grab and swallow" method. Accommodation and care are as for the Racer and Whip Snakes. Feed on mice, rats, or chickens and provide facilities to climb and bathe.

Corn Snake
Elaphe guttata

Perhaps one of the best loved of all North American serpents, the Corn Snake or Red Rat Snake is now being bred selectively for color variations, though it would be hard to better the original color pattern of the wild snake, which is reddish buff with a series of large black-bordered reddish blotches running along the body. A subspecies known as the Great Plains Rat Snake, *E. g. emoryi*, is brown to grayish with darker brown or gray blotches. Reaching a maximum length of about 180 cm (6 ft), the species is found from the southeastern USA to the Great Plains and south into Mexico. Its habitat is varied from woodland to open grassland and cultivated land, where it is often encouraged due to its rodent-consuming activities. Mainly crepuscular and terrestrial, it can climb well should the occasion arise and often hides in the burrows of rodents. In captivity it requires a medium-sized, mildly heated

The incredible Eastern Indigo Snake, *Drymarchon corais couperi*, is one of the most reliable of all captive reptiles. Unfortunately, it is also severely endangered. Photo by R. D. Bartlett.

Rat Snake
Elaphe obsoleta.

The North American Rat Snakes form a complex group of subspecies and intergrades. The best known subspecies are probably the Yellow Rat Snake, *E. o. quadrivittata*, and the Texas Rat Snake, *E. o. lindheimeri*. The former is yellow to olive-yellow with darker lines running the length of the body; the latter is blackish with buff blotches. Other subspecies include the Black, *E. o. obsoleta*; Everglades, *E. o. rossalleni*; and Gray, *E. o. spiloides*. Most average about 150 cm (5 ft) in length. Taking kindly to captivity, all members of the group are frequently kept in the terrarium and many color varieties are being bred. However, although this is a very interesting aspect of snake keeping and there is certainly nothing wrong with breeding color varieties, it is highly recommended that pure strains of the natural terrarium with facilities for climbing and bathing. Feed on a diet of mice. Allow to hibernate in winter for a few weeks.

A very popular hobbyist choice is the beautiful and hardy Corn Snake, *Elaphe guttata guttata*. Photo by Isabelle Francais

Note the difference between the normal (below) and albinistic varieties (above) of the Corn Snake. Photos by Louis Porras.

subspecies continue to be maintained lest they be lost forever! Native to North America from the Great Lakes to Texas and into Mexico, Rat Snakes are found in a great range of habitats including farmland, where they are welcome as rodent exterminators.

Another popular *Elaphe* is the Yellow Rat Snake, *Elaphe obsoleta quadrivittata*. Photo by Louis Porras.

Housing and care are as described for the Corn Snake, with temperatures reflecting the natural habitat.

There are many other North American, European, and Asian species in this large genus, too many to cover in detail here. All require similar care. These include the Aesculapean Snake, *E. longissima* (Europe); Mandarin Snake, *E. mandarinus* (Asia); Four-lined Snake, *E. quatuorlineata* (Europe); Leopard Snake, *E. situla* (Europe); Trans-Pecos Rat Snake, *E. subocularis* (Texas to Mexico, now in genus *Bogertophis*); Beauty Snake, *E. taeniura*; and Fox Snake, *E. vulpina* (Great Lakes region of North America).

Common Kingsnake
Lampropeltis getula

Like the Rat Snakes, the Common Kingsnake also is an extremely popular terrarium reptile and also is a subspecies complex. The best known subspecies perhaps are the California Kingsnake, *L. g. californiae*, which usually has white bands on black, but a striped form exists; the Florida Kingsnake, *L. g. floridana*, which is brownish black with yellow markings on the scales; the Eastern Kingsnake, *L. g. getula*, with similar white and black bands to the California King but the white bands join at the base, making a chain-like pattern; and the Speckled Kingsnake, *L. g. holbrooki*, which is glossy black with numerous small yellow spots. Several subspecies, especially the California, are readily

California Kingsnakes, *Lampropeltis getula californiae*, come in two main pattern varieties: striped and, like the one shown below, banded. Photo by F. J. Dodd.

SQUAMATA

Without a doubt, one of the most stunning of all *Lampropeltis* is the Sinaloan Milk Snake, *Lampropeltis triangulum sinaloae*. This snake has been bred repeatedly in captivity over the last few years (thanks to popular demand) and is now available many places snakes are sold. Photo by K. T. Nemuras.

available as albinos. Kingsnakes range throughout the southern half of the USA and into Mexico. There is a considerable variation in habitat from woodland to semi-desert. The wild diet includes other snakes, even rattlesnakes, which probably accounted for the species's common name. Kingsnakes require a medium-sized terrarium with facilities to climb. A hollow log or hide box should be provided as well as facilities to bathe. Local temperatures to 28°C (82°F) with cooler retreats are best. Turn off the heat at night. Allow a short period of winter hibernation. Feed on small rodents and chicks, especially pinkie mice. *Lampropeltis alterna*, the Gray-banded Kingsnake, is heavily bred in captivity but expensive. It is more nocturnal than the Common Kingsnake and can take drier and warmer surroundings.

Milk Snake
Lampropeltis triangulum

This species gets its common name from an old wives' tale that it sucks milk from the udders of cows. It is a handsome species, much in demand for the terrarium. Reaching a length of 110 cm (44 in), often larger in tropical subspecies, it is marked in broad red and narrower yellow or cream bands separated by black bands. There are many subspecies found throughout the eastern half of the USA, Mexico, and Central American to northwestern South America. It may be active day or night, usually preferring open woodlands. A shy and secretive snake, it spends most of its time under ground litter or in burrows where it feeds mainly on lizards, small

Another commonly seen member of the *triangulum* species is the Eastern Milk Snake, *L. t. triangulum*. Photo by R. T. Zappalorti.

Although referred to as a kingsnake, the Scarlet Kingsnake, *Lampropeltis triangulum elapsoides*, is actually a milk snake subspecies. Photo by R. T. Zappalorti.

rodents, and invertebrates. It can be trained to eat pinkie mice in captivity. The terrarium should have ample controllable hiding refuges.

Viperine Snake
Natrix maura

The common name of this snake is derived from its somewhat viperine shape, though it is in fact quite harmless (it is frequently mistaken by people for one of the European vipers, an unfortunate fact that often accounts for its violent end). Averaging 100 cm (39 in) when adult, its head is large and triangular and the body short and robust. The ground color is olive brown to green. There are two rows of staggered dark blotches down the back that sometimes form a zig-zag pattern. Native to southwestern Europe and North Africa, it is nearly always found in or near water. It often basks at the water's edge and dives for cover if alarmed. A diurnal species that is irritable and nervous at capture, it soon settles into life in the terrarium. It requires an aqua-terrarium with the water temperature maintained around 24°C (75°F) and an air temperature reaching 30°C (86°F), reduced at night. A period of winter hibernation is recommended. Feed on fish, frogs, and earthworms. Other European species in the same genus and requiring similar care include the Grass Snake, *N. natrix*, and the Dice Snake, *N. tessellata*.

Even though their popularity seems to shift, for the most part water snakes, like this *Natrix maura*, do not make good captives. Photo by E. Zimmermann.

Mississippi Green Water Snake
Nerodia cyclopion

There are several North American water snakes in this genus, and *Nerodia cyclopion* is fairly typical of the group. (North American water snakes formerly were in the genus *Natrix* but were split off a decade or so ago.) In spite of its common name, the Mississippi Green Water Snake may be brown, reddish, or olive green in color. It has indistinct black bars along the flanks alternating with crossbars on the back; juveniles show the markings more plainly. The underside is brown with pale halfmoon spots. Reaching 180 cm (72 in), but usually shorter, it is a heavily built snake usually inhabiting swamps, ditches, and bayous in the Mississippi River valley of the USA. It is primarily diurnal and feeds largely on fishes. Care of this and other *Nerodia* species is similar to that described for *Natrix maura*. Water snakes cannot be kept in water as they will come down with blister disease and die.

Pine, Bull, and Gopher Snakes
Pituophis melanoleucus

Reaching a length of 250 cm (100 in), this is one of the more popular terrarium snakes. It is a robust species that occurs in a multitude of subspecies grouped into three types (Pine, *melanoleucus*; Bull, *sayi*; Gopher, *catenifer*) that probably represent three full species. The Northern Pine Snake, *P. m. melanoleucus*, is creamy white in color, with a row of large black, brown, or reddish brown blotches along the back and smaller ones on the sides. Other subspecies are similar but differ in color and to some extent pattern, including entirely black

Facing page: Water snakes can give bites that may take time to stop bleeding. This is due to an anticoagulant in their saliva. Photo of *Nerodia floridana* by R. T. Zappalorti.

This is a photo of a very rare snake indeed: it is a "snow" albino Northern Pine Snake, *Pituophis melanoleucus melanoleucus*. Photo by R. T. Zappalorti.

forms and striped forms. The species (in the broad sense) ranges over most of the USA and into Mexico. The Pine Snake is mainly diurnal, but may be active at night in warmer weather. It often takes refuge in the burrows of other animals and is an avid consumer of small mammals. When disturbed, a wild specimen hisses loudly, vibrates its tail, and lunges at the intruder. In captivity it soon settles and doesn't mind being handled. Feed on mice or rats. It requires a large terrarium with facilities to climb and a temperature around 25°C (77°F) with warmer basking areas. Reduce the temperature at night and allow to hibernate in winter.

Oriental Rat Snake or Dhaman
Ptyas mucosus

The Sanskrit name of this species, Dhaman, actually means "a rope," which probably arose from its habit of striking boldly upward like a whipcord when it is provoked. It is a powerfully built snake and one of the largest colubrids, with a maximum length of 3.6 m (12 ft). It is usually dull brown to olive, with faint spots on the scales starting in the rear half of the body and forming faint diagonal bands. Native to central, southern, and southeastern Asia, it occurs mainly in areas of dry scrubland and open forest. Mainly diurnal, it feeds on a variety of small mammals, birds, and reptiles. Though fierce and aggressive when captured, it soon settles into the terrarium, which should be roomy and have facilities for it to bathe and climb. Local basking areas should reach about 30°C (86°F), but the temperature should be reduced at night. A period of hibernation at reduced temperatures is recommended. This species will do well on a diet of mice, rats, and chickens. The closely related Chinese

The Oriental Rat Snake, *Ptyas mucosus*, is also known as the "Dhaman," which means "rope." Photo by C. Banks.

Rat Snake, *P. korros*, is somewhat shorter in total length and requires similar care in captivity.

Chicken Snake
Spilotes pullatus

Growing to a maximum of 10 feet, this is a graceful snake with a relatively small head and a long, tapered tail. The body is flattened laterally and is patterned in brilliant yellow and black. Native to Central and South America it occurs in at least 5 subspecies. Semi-arboreal and diurnal, it is usually found in wooded country, often close to water. Although it is very irritable when first captured, it soon settles into the terrarium and can become quite tame. It requires a large, tall, terrarium with several stout climbing branches. Air temperature may reach 30°C (86°F) during the day, but should be reduced to around 22°C (71°F) at night. A short winter rest period at reduced temperatures is also recommended. Feed on mice, rats, and chickens of a suitable size and provide facilities for bathing.

Common Garter Snake
Thamnophis sirtalis

Being small, docile, attractive, easy to feed, and easily obtainable, garter snakes of several species are very popular terrarium subjects, and many a budding herpetologist has started off with these charming little reptiles. The Common Garter Snake averages about 60 cm (24 in), but specimens in excess of 130 cm (50 in) have been recorded. There are several subspecies, but average coloration is brown to olive-brown with three yellow stripes running along the back and sides and a series of dark, angular blotches running between the stripes. Found over most of the USA except for the far Southwest, this species lives in damp areas never far from water, which it will enter in search of tadpoles, frogs, and small fish. It requires an aqua-terrarium with dry basking areas. Provide heat to 25°C (77°F) with a warmer basking site. Reduce the temperature at night and allow to hibernate in winter. It will feed on earthworms and small fish, as well as frogs.

Melanism is rare in most species, but in Common Garter Snakes, like this *Thamnophis sirtalis sirtalis*, it is not so unusual. Photo by R. Allan Winstel.

Venomous Snakes

As they are certainly not recommended for the home terrarium keeper, especially the beginner, venomous snakes will not be dealt with in detail here. However, most terrarium keepers are interested in all snakes, so we will undertake a brief survey of the venomous snake groups.

REAR-FANGED SNAKES

Not all members of the family Colubridae are totally harmless, and there are several subfamilies (notably Boiginae) that contain species of rear- or back-fanged snakes. While the bite or venom of most of these has never been proven to be dangerous to man, there are a few species in which the venom has caused severe illness or even death. The term rear-fanged arises from the fact that the elongate venom-carrying teeth are set not in the front of the upper jaw as in all other venomous snakes, but well back in either side of the upper jaw. The venom glands are situated in the cheeks above the jaw, and when the snake bites to subdue its prey or in defense, venom is released along a groove in one pair or more of these fangs and thus passed into the victim. Most of these species consume relatively small prey that they have no problem envenomating, but if they should strike a person in defense they rarely are able to open the mouth wide enough to bring the rear fangs into play. There are some exceptions to this rule, however, notable ones being the Boomslang (the word arises from the Dutch for "tree snake"), *Dispholidus typus*, and the African Bird or Vine Snake, *Thelotornis kirtlandi*, both of which have caused several human deaths, including the death of one well-known herpetologist each! The venom of these snakes is known to be particularly virulent but, to make things a little worse, the rear-fangs in these two particular species are set further forward in the jaw than in most other rear-fanged snakes. It pays never to be complacent with any rear-fanged snakes; even those that are classed as "mildly venomous" have, on occasion, given a nasty surprise to their trusting handler!

FAMILY ELAPIDAE

COBRAS, MAMBAS, KRAITS, AND CORAL SNAKES

The Elapidae contains some 50 genera and about 200 species of venomous snakes distributed throughout the tropics and subtropics. However, the headquarters of the family must be Australia, where elapids are the ruling group of snakes. Australia has the distinction of being the only country where venomous snake species outnumber non-venomous ones. As

Visually reminiscent of the California Kingsnake, this Bandy-Bandy, *Vermicella annulata*, is actually deadly venomous. Photo by C. Banks.

Australian elapids do not fit easily into the groups comprising the cobras, mambas, kraits, and coral snakes, it has recently been suggested that the majority of Australian elapids should be assigned to the family Hydrophiidae, the sea snakes, with which they have some similarities. Lengths of most elapid species are in the range of 30-100 cm (12-40 in); notable exceptions are the King Cobra, *Ophiophagus hannah*, of Southeast Asia with a record length of 5.6 m (18.2 ft), making it the world's longest venomous snake, and the Taipan, *Oxyuranus scutellatus*, of Australia with a record length of 4 m (13 ft). Most of the species are slender and colubrid-like in form with rather short snouts and glossy scales, but there are many exceptions (for example the Australian Death Adders, *Acanthophis*, which superficially resemble viperids).

All elapid species possess a pair of fixed, hollow fangs at the front of the upper jaw that are used to inject venom into their prey from the large modified salivary glands in the cheeks. These fangs are relatively short when compared with those of the Viperidae and Crotalidae, but nevertheless very effective. The venom of almost all elapids is mainly neurotoxic (attacking the nervous system), with only a minimal hemotoxic (attacking the blood and tissues) effect in most cases. Many elapid species are highly dangerous to humans, and no species in the group is recommended for the home terrarium keeper. They are best left to the attentions of the zoo keeper or professional herpetologist.

Facing page: The Common Death Adder, *Acanthophis antarcticus*, is found in most parts of Australia and is primarily nocturnal. Photo by C. Banks.

It would not be advisable for an amateur to keep something as dangerous as this Cape Coral Snake, *Aspidelaps lubricus*, since it can kill a human being easily. Photo by K. H. Switak.

Elapid species are found in a variety of habitats. There are those that are primarily terrestrial (including burrowing types), those that are mainly arboreal, and a few that are semi-aquatic. They may be oviparous (egg-laying) or ovoviviparous (giving birth to live young).

FAMILY HYDROPHIIDAE

SEA SNAKES

Perhaps the least studied of snake groups until very recently, the family contains between 50 and 75 species of almost totally marine snakes that inhabit the Pacific and Indian Oceans and adjacent tropical and subtropical seas from whence they sometimes accidentally find their way into temperate waters. One or two species have become endemic to freshwater lakes. The longest sea snakes may be 275 cm (8.5 ft) long. Most are strongly adapted to a

The Australian Desert Death Adder, *Acanthophis pyrrhus*, has strongly keeled dorsal scales. Photo by C. Banks.

Due to their obviously highly demanding captive needs, sea snakes like this Pelagic Sea Snake, *Pelamis platurus*, are almost never kept domestically. Photo by Paul Freed.

Another aspect of the sea snake family that makes them poor captives is their fatally toxic venom. Photo of the Yellow-lipped Sea Snake, *Laticauda colubrina*, a particularly lethal species, by R. E. Kuntz.

marine existence, some having a laterally flattened body and a rudder-like tail to aid in swimming. As with most aquatic snakes, the eyes are relatively small and the nostrils are valved and kept closed when the reptiles are submerged. Sea snakes possess a gland in the floor of the mouth surrounding the tongue sheath that removes excessive salt from the blood stream, the salt being expelled when the tongue is extended. Like members of the Elapidae (to which they

It is perhaps unfortunate that such graceful snakes as the sea kraits, *Laticauda*, cannot be maintained domestically. Photo by Dwayne Reed.

are very closely related), sea snakes possess short venom fangs at the front of the upper jaw followed by varying numbers of smaller, backwardly directed teeth on the jaws and sometimes on the palate. Sea snakes feed largely on fish, especially eels. Almost all sea snake species spend their whole lives in the water, giving birth to live young, but members of one genus (*Laticauda*) have to go ashore to lay their eggs,

This is the dark phase of the Northern Adder, *Vipera berus,* an interesting species in that it is the most northerly ranging serpent in the world. Photo by S. Kochetov.

which are usually buried in sand above the high water mark. (Most herpetologists today would consider *Laticauda* to be an elapid, not a sea snake.) Sea snakes are rarely kept in captivity and their husbandry is poorly researched, though some public aquaria are now regularly exhibiting them in huge seawater tanks and marine biologists are making further investigations into their biology. The venom is highly toxic to humans, although most species seem reluctant to bite once out of the water.

FAMILY VIPERIDAE

TYPICAL OR OLD WORLD VIPERS

The true vipers are a family containing just 50 or 60 species in Europe, Asia, and Africa in tropical to temperate climates. One species (*Vipera berus*, the Northern Adder) reaches within the Arctic Circle in Europe; such northerly specimens are only active for 4-5 months in the year. Many species are adapted to desert conditions, but others are found in savannah to tropical rain forest. Numerous species are fairly nocturnal but may also spend much of the day basking in the sun. They possess a pair of relatively

Below: The Copperhead, *Agkistrodon contortix*, is one of many venomous snakes known as a "solenoglyph." Photo by S. Kochetov. *Opposite:* "Solenoglyphs" are snakes that have movable front fangs, like those shown on this Gaboon Viper, *Bitis gabonica*. Photo by John Visser.

The Cottonmouth, *Agkistrodon piscivorus*, is extremely dangerous and does not take well to either captivity or human company. Photo by R. T. Zappalorti

long, hollow venom fangs that rest along the roof of the mouth and are brought forward by rotation of the maxilla as the reptile strikes. In most cases the prey is released after envenomation and dies in a few minutes; the snake follows its scent trail and devours it. Better known members of the family include the huge, colorful Gaboon Viper, *Bitis gabonica*, which can reach 180 cm (6 ft) in length; and the Puff Adder, *B. arietans*. All species in the family are highly venomous and dangerous to man, mostly producing hemotoxic (attacking the blood) venom. They should only be kept by professional or highly experienced herpetologists. Extremely strict safety precautions are required when such species are kept in captivity.

FAMILY CROTALIDAE

PIT VIPERS

This family contains almost 150 species mainly native to the Americas, but with some representatives in southern and southeastern Asia. (Many herpetologists consider the pit vipers to be members of the family Viperidae, not recognizing the Crotalidae as a good family.) They are found in a range of climates from cool to warm temperate, subtropical, and tropical. Lengths vary from 60-350 cm (24-140 in). The large, often triangular, head is separated from the robust body by a relatively narrow neck. Rattlesnakes, moccasins, copperheads, fer-de-lances, and tree vipers are representative types. The more primitive types (*Agkistrodon, Sistrurus*) possess large, symmetrical head shields, but the majority have small or irregular head scales. The main characteristic of the family that separates them from the Viperidae is the pit or thermoreceptor situated on each side of the head between the eye and the nostril. This organ senses infra-red (heat) radiation that is used to locate warm-blooded prey in conditions of poor visibility. Like the viperids, crotalids have large, hinged venom fangs in the front of the upper jaw. Most of the species are highly venomous, and comments regarding the dangers of keeping viperids also apply to members of this group.

INDEX

NOTE: **Boldface** numbers indicate illustrations

Acanthophis antarcticus, **241**
Acanthophis pyrrhus, **243**
Acquiring specimens, 67-68
Acrochordus javanicus, 200
African Rock Python, 212
Agama agama, **156**, 157-158
Agkistrodon contortix, **248**
Agkistrodon piscivorus, **250**
Algerian Sand Racer, **177**, 178
Alligator mississippiensis, **141**
Ameiva ameiva, 179, **180-181**
American blind snakes, 196
Amphibolurus barbatus, 158-159, **159**
Anguis fragilis, 182-183
Anolis carolinensis, **24, 105,** 152-154
Arizona sp., **37**
Asian Water Dragon, 159-160, **161**
Aspidelaps lubricus, **242**
Australian Death Adder, **243**
Australian Snake-necked Turtle, 138
Ball Python, **210**, 211-212
Bandy-Bandy, **239**
Basiliscus plumifrons, **153**
Basiliscus sp., 154
Basilisks, 154
Beaked Blind Snake, **197**
Bearded Lizard, 158-159, **159**
Binomial nomenclature, 32
Birth, live, 112-114
Bitis gabonica, **249**
Black Rat Snake, **89**
Blind snakes, 196, **197**
Boa Constrictor, 201-203, **202**
Bog Turtle, **19, 121**

Box Turtle, **132-133**
Branches, 57
Breeding colonies, 104
Broadhead Skink, **18**
Bullsnake, 231-234
California Kingsnake, **223**
Canary Island Lizard, 174-175
Cape Coral Snake, **242**
Carnivorous, definition, 72
Central Plains Milk Snake, **106**
Chalcides ocellatus, 168-169, **168**
Chameleo jacksoni, 162-165, **164**
Chameleo pardalis, 162, **163**
Chelodina longicollis, 138
Chelus fimbriatus, **139**
Chelydra serpentina, 127-128, **129**
Chicken Snake, 235-236
Chickens, 86
Chit-Chat, 148-150
Chondropython viridis, 208-209, **209**
Chrysemys picta, 132
Classificiation, charts, 36
Cleaning, 90-92
Clemmys muhlenbergii, **19, 121**
Cnemidophorus sexlineatus, 179-180
Cobras, 239-243
Coleonyx variegatus, 145
Collared Lizard, **34**
Coluber constrictor priapus, **215**
Coluber constrictor, **35,** 215-217
Common Garter Snake, 236, *melanistic* variety, **237**
Common Iguana, **102, 155,** 155-157
Common Snapping Turtle, 127-

INDEX

128, **129**
Common Tegu, **181**, 181-182
Conservation, 18-26
Copperhead, **248**
Coral snakes, 239-243
Cordylus giganteus, **171**, 172
Corn Snake, **118**, 218-220, **220**, **221**
Coronella austriaca, **216**, 217
Corucia zebrata, 170
Cottonmouth, **250**
Crickets, 79-81
Crocodile Lizard, 186-188
Crocodylus niloticus, 140
Crocodylus porosus, 140
Crotaphytus collaris, **34**
Cuora sp., 134-135
Cylindrophis rufus, 198-199
Day gecko, **11**, 150-151
Death Adder, **241**
Desert Banded Gecko, **146-147**
Desert Iguana, 154-155
Desert Tortoise, *albino*, **122**
Dhaman, 234-235, **235**
Dipsosaurus dorsalis, 154-155
Drymarchon corais couperi, **219**
Drymarchon corais, 217-218
Earthworms, 83-84
Eastern Blue-tongued Skink, 169-170
Eastern Box Turtle, 133-134
Eastern Hognose Snake, **28**
Eastern Indigo Snake, **219**
Eastern Milk Snake, **113**, **227**
Eastern Mud Turtle, 128
Ectoparasites, 94-96
Egg tooth, 119
Egglaying, 112-114
Egglaying, box, 113-114
Egglaying, sites, 112
Eggs, evolutionary development of reptiles', 27-29
Elaphe guttata, **118**
Elaphe guttata, 218-220, **220**, **221**
Elaphe obsoleta obsoleta, **89**
Elaphe obsoleta quadrivittata, **222**
Elaphe obsoleta, 220-222
Endoparasites, 96-98
Epicrates cenchria, 203-204, **203**
Eryx conicus, 207, **208**
Estuarine Crocodile, 140
Eublepharus macularius, 146-147, **147**
Eumeces fasciatus, 165-168, **166-167**
Eumeces laticeps, **18**
Eunectes murinus, 204-206, **205**
Eyed Skink, 168-169, **168**
Feeding strategies, 88-89
Five-lined Skink, 165-168, **166-167**
Flat Rock Lizard, 172-173
Flies, 81-83
Florida Green Water Snake, **230**
Florida Soft-shell, 137, **136**
Florida Worm Lizard, 194
Foods, processed, 87-88
Gaboon Viper, 249
Gadow, Hans, 8
Gallotia galloti, 174-175
Garter snake, **10**
Gekko gecko, 147-148, **149**
Geochelone carbonaria, **107**, 127
Geochelone denticulata, **91**
Geochelone elegans, **125**, 127
Geochelone pardalis, 126-127
Geochelone radiata, **111**
Georgia Blind Salamander, **21**
Gerrhonotus multicarinatus, 185-186, **187**
Gerrhosaurus major, 173-174
Gestation, 112-113
Giant Sungazer, **171**, 172
Gila Monster, **192**
Glossy snake, **37**
Gopher snakes, 231-234
Gopher Tortoise, 126
Gopherus agassizi, **122**
Gopherus polyphemus, 126
Gray-banded Kingsnake, **98-99**
Greek Tortoise, 123-126
Green Anaconda, 204-206, **205**

INDEX

Green Anole, **24**, 152-154, **105**
Green Basilisk, **153**
Green Lizard, **175**, 175-177
Green Tree Python, 208-209, **209**
Greenhouses, 51
Habitat destruction, 22
Haemorrhois viridiflavus, 214-215
Haideotriton wallacei, **21**
Handling, 69-71
Heaters, aquarium-type, 61-62
Heaters, cable, 62
Heaters, pad, 62
Helmeted Turtle, 138
Heloderma suspectum, **192**
Hemidactylus frenatus, 148-150
Herbivorous, definition, 72
Heterodon platirhinos, **28**
Hibernation, 92-93
"Hot Rocks", 62
Humidity, 64-66
Iguana iguana, **102, 155**, 155-157
Incubation, 114-120
Indian Python, 209-211
Indigo snake, 217-218
Infections, bacterial, 98
Infections, protozoan, 98-100
Infections, respiratory, 101-102
International Code of Zooligical Nomenclature, 37
Jackson's Chameleon, 162-165, **164**
Javan Wart Snake, 200
Jordan's Salamander, **26**
Jungle Runner, **179**, 180-181
Kingsnake, 222-228
Kinosternon subrubrum hippocrepis, **130**
Kinosternon subrubrum subrubrum, 128
Kraits, 239-243
Lacerta viridis, **175**, 175-177
Lampropeltis alterna, **98-99**
Lampropeltis getula californiae, **223**
Lampropeltis getula, 222-228
Lampropeltis pyromelana pyromelana, **31**
Lampropeltis triangulum elapsoides, **228**
Lampropeltis triangulum gentilis, **106**
Lampropeltis triangulum sinaloae, **114-115, 224-225**
Lampropeltis triangulum triangulum, **113, 227**
Lampropeltis triangulum, 226-228
Laticauda colubrina, **245**
Laticauda sp., **246**
Leopard Gecko, 146-147, **147**
Leopard Tortoise, 126-127
Lichanura trivirgata, 206-207
Lighting, 62-64
Linne, Karl von, (Linnaeus), 32
Liotyphlops albirostris, **198**
Livefoods, collecting, 75-77
Livefoods, culturing, 77-84
Livefoods, invertebrate, 84
Livefoods, vertebrate, 86-87
Locusts, 81
Loggerhead Musk Turtle, **39**
Logs, 57
Madagascar Day Gecko, **150, 151**
Malaclemys terrapin rhizophorarum, **17**
Mambas, 239-243
Mangrove Diamondback Terrapin, **17**
Matamata, 138-139, **139**
Mealworms, 77-78, **78**
Mice, 84-86
Milk Snake, 226-228
Mississippi Green Water Snake, 231
Mississippi Mud Turtle, **130**
Mites, 94-96
Musk Turtle, 128-130
Natrix maura, 228-231, **229**
Nerodia cyclopion, 231
Nerodia floridana, **230**
Nile Crocodile, 140
Northen Pine Snake, **20**
Northern Adder, **247**

INDEX

Northern Pine Snake, "snow albino", **232-233**
Nutrition, 71-72
Nutritional problems, 93
Omnivorous, definition, 72
Ophisaurus apodus, 183-185, **184**
Ophisaurus attenuatus, 185
Oriental Rat Snake, 234-235, **235**
Outdoor setups, 47-51
Painted Turtle, 132
Panther Chameleon, 162, **163**
Parthenogenic, definition, 108-109
Pelagic Sea Snake, **244**
Pelamis platurus, **244**
Pelomedusa subrufa, 138
Phelsuma madagascariensis, **150**, **151**
Phelsuma sp., 150-151
Phelusuma astriata, **11**
Physignathus cocincinus, 159-160, **161**
Pine snakes, 231-234
Pit vipers, 25
Pituophis melanoleucus melanoleucus, **20**, **232-233**
Pituophis melanoleucus, 231-234
Plants, 57-59
Platysaurus guttatus, 172-173
Plethodon jordani, **26**
Podarcis muralis, **176**, 177-178
Pollution, 22-24
Probe, sexing, **108-109**, 109-110
Psammodromus algirus, **177**, 178
Pseudemys scripta elegans, **25**, **33**, 131
Ptyas mucosus, 234-235, **235**
Python molurus, 209-211
Python regius, **210**, 211-212
Python reticulatus, **101**, 212-213, **213**
Python sebae, 212
Quarantine, 69
Racer, **35**, 215-217
Radiated Tortoise, **111**
Rainbow Boa, 203-204, **203**

Rainbow Lizard, **156**, 157-158
Rat snake, 220-222
Rats, 84-86
Rear-fanged snakes, 238
Rearing, young, 120-120
Red Cylinder Snake, 198-199
Red-eared Slider, **25**, **33**, 131
Red-footed Tortoise, **107**, 127
Reproductive cycle, 103-108
Reptile, definition, 7
Reptiles, evoultion, 15, 27-31
Reptiles, over-collecting and trading of, 24-25
Reticulated Python, **101**, 212-213, **213**
Rhineura floridana, 194
Rhinotyphlops schinzi, **197**
Rhynchosauria, 30
Rocks, 52-57
Rosy Boa, 206-207
Rough-tailed Sand Boa, 207, **208**
Roundworms, 96-98
Safety precautions, housing, 66
San Francisco Garter Snake, **23**
Savannah Monitor, 190, **191**
Scarlet Kingsnake, **228**
Sea krait, **246**
Sea snakes, 243-247
Sex determination, 108-112
Sheltopusik, 183-185, **184**
Shield-tail snakes, 199
Shinisaurus crocodilurus, 186-188
Sinaloan Milk Snake, **114-115**, **224-225**
Six-lined Racerunner, 179-180
Skin problems, 100-101
Slender Glass Lizard, 185
Slow Worm, 182-183
Smooth Snake, **216**, 217
Soft-shelled Turtles, 135
Solomons Giant Skink, 170
Sonoran Mountain Kingsnake, **31**
Southern Alligator Lizard, 185-186, **187**
Southern Black Racer, **215**
Species, introduced, 25-26

INDEX

Sphenodon punctatus, 30, **30, 142-143**, 142-143
Spilotes pullatus, 235-236
Spotlamps, 60-61
Spur-thighed Tortoise, 123-126
Starred Tortoise, **125**, 127
Sternotherus minor, **39**
Sternotherus odoratus, 128-130
Stinkpot, 128-130
Substrate materials, 52
Sudan Plated Lizard, 173-174
Tapeworms, 96-98
Temperature, controlling, 59-60
Tenebrio molitor, **78**
Terrapene carolina, 133-134, **132-133**
Terraria, built-in, 45-47
Terraria, glass, 44-45
Terraria, sizes, 41-43
Terraria, types of, 40-41
Terraria, wooden, 43-44
Testudo graeca, 123-126
Thamnophis sirtalis sirtalis, 236, **237**
Thamnophis sirtalis tetrataenia, **23**
Thamnophis sp., **10**
Thread blind snakes, 196-197
Ticks, 94-96
Tiliqua scincoides, 169-170
Tokay Gecko, 147-148, **149**
Transport of specimens, 68-69
Trionyx ferox, **136**
Tuatara, 30, **30, 142-143**, 142-143
Tungsten (incandescent) lamps, 60
Tupinambis teguexin, **181**, 181-182
Varanus exanthematicus, 190, **191**
Varanus salvator, 189-190
Vegetables, 87
Vermicella annulata, **239**
Vermiculite, 116
Vipera berus, **247**
Viperine Snake, 228-231, **229**
Vipers, 247-251
Wall Lizard, **176**, 177-178
Water Monitor, **189**, 189-190
Western Banded Gecko, 145
Western Whip Snake, 214-215
Wounds, 93-94
Yellow Rat Snake, **222**
Yellow-footed Tortoise, **91**
Yellow-lipped Sea Snake, **245**